McGraw-Hill
My Math

This is your very own math book! You can write in it, draw, circle, and color as you explore the exciting world of math.

Let's get started. Grab a crayon and draw a picture that shows what math means to you.

Have fun!

This is your space to draw.

Mc
Graw
Hill
Education

MHEonline.com

STEM McGraw-Hill is committed to providing
instructional materials in Science, Technology, Engineering,
and Mathematics (STEM) that give all students a solid
foundation, one that prepares them for college and careers
in the 21st century.

Send all inquiries to:
McGraw-Hill Education
8787 Orion Place
Columbus, OH 43240

ISBN: 978-0-07-668790-9 (*Volume 1*)
MHID: 0-07-668790-2

Printed in the United States of America.

7 8 9 10 11 LMN 23 22 21 20 19

Understanding by Design® is a registered trademark of the Association for Supervision
and Curriculum Development ("ASCD").

Meet The Artists!

Alyssa Gonzalez

King of the Math Jungle This was a great experience similar to being on a roller coaster. Many friends and family supported me in this competition and many kids at Veterans Elementary will benefit from this. *Volume 1*

Finley Moss

Math is Yummy I came up with the idea because my mom and I love to bake cakes and it takes both time and measurement to do so. I was dreaming about how it would feel to win, so I am very excited! Math is yummy after all! *Volume 2*

Other Finalists

Landre Kate Beeler
Pop Numbers 4 Me

Sherry Bergeron's Class*
Math Makes Our World a Better Place

Sally Barmakian's Class*
Math is a Rainbow of Learning

Kamya Cooperwood
Monkey

Judson Upchurch
I like Numbers

Charles "Greyson" Biggs
Rainbow of Possibilities

Leah Rauch
Connect the Dots

Andrew Morris
Math Designed World

Matthew Saldivat
Math Guy in the Sky

Ngun Za Cin
Three-Dimensional Shape Sculpture

Find out more about the winners and other finalists at www.MHEonline.com.

We wish to congratulate all of the entries in the 2011 *McGraw-Hill My Math* "What Math Means To Me" cover art contest. With over 2,400 entries and more than 20,000 community votes cast, the names mentioned above represent the two winners and ten finalists for this grade.

* Please visit mhmymath.com for a complete list of students who contributed to this artwork.

GO digital

it's all at
connectED.mcgraw-hill.com

Go to the Student Center for your eBook, Resources, Homework, and Messages.

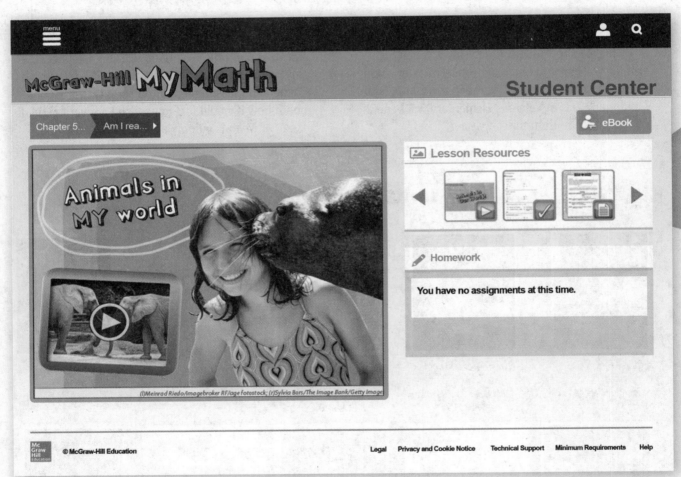

Write your Username [   ]   Password [   ]

Get your resources online to help you in class and at home.

Vocab

Find activities for building vocabulary.

Watch

Watch animations of key concepts.

Tools

Explore concepts with virtual manipulatives.

eHelp

Get targeted homework help.

Games

Reinforce with games and apps.

Tutor

See a teacher illustrate examples and problems.

GO mobile

Scan this QR code with your smart phone* or visit mheonline.com/stem_apps.

*May require quick response code reader app.

Available on the App Store

Contents in Brief
Organized by Domain

Processes & Practices → Woven Throughout

Chapter 1
Apply Addition and Subtraction Concepts

ESSENTIAL QUESTION
What strategies can I use to add and subtract?

Getting Started

I ♥ animals!

Lessons and Homework

Wrap Up

connectED.mcgraw-hill.com

Watch ▶ **Look for this!** Click online and you can watch videos that will help you learn the lessons.

Chapter 2 — Number Patterns

ESSENTIAL QUESTION
How can equal groups help me add?

Getting Started

Lessons and Homework

Wrap Up

Welcome to the desert, cowpoke!

Hi!

connectED.mcgraw-hill.com

Chapter 3 Add Two-Digit Numbers

Getting Started

Lessons and Homework

Wrap Up

We make a great pair!

connectED.mcgraw-hill.com

eHelp
Look for this!
Click online and you can get more help while doing your homework.

Chapter 4 Subtract Two-Digit Numbers

ESSENTIAL QUESTION
How can I subtract
two-digit numbers?

Getting Started

Lessons and Homework

Wrap Up

Springtime!

connectED.mcgraw-hill.com

Chapter 5 Place Value to 1,000

Getting Started

Lessons and Homework

Wrap Up

Look how I've grown!

connectED.mcgraw-hill.com

 Tools

Look for this!
Click online and you can find tools that will help you explore concepts.

Chapter 6 Add Three-Digit Numbers

Getting Started

Lessons and Homework

Wrap Up

connectED.mcgraw-hill.com

My house

Chapter 7 — Subtract Three-Digit Numbers

Getting Started

Lessons and Homework

Wrap Up

Look for this!
Click online and you can find activities to help build your vocabulary.

You + School = Cool

connectED.mcgraw-hill.com

Chapter 8 Money

ESSENTIAL QUESTION
How do I count and use money?

Getting Started

Lessons and Homework

Wrap Up

You make me laugh!

Check
Look for this!
Click online and you can check your progress.

connectED.mcgraw-hill.com

Chapter 9 Data Analysis

ESSENTIAL QUESTION
How can I record and analyze data?

Getting Started

Go nuts!

Lessons and Homework

Wrap Up

connectED.mcgraw-hill.com

Chapter

10 Time

ESSENTIAL QUESTION
How do I use and tell time?

Getting Started

Lessons and Homework

Wrap Up

It's time for our trip!

connectED.mcgraw-hill.com

Chapter 11

Customary and Metric Lengths

Measurement and Data

ESSENTIAL QUESTION
How can I measure objects?

> I love rulers because I'm an inch worm!

Getting Started

Lessons and Homework

Wrap Up

connectED.mcgraw-hill.com

Chapter 12

Geometric Shapes and Equal Shares

ESSENTIAL QUESTION
How do I use shapes and equal parts?

Come on guys. Let's check it out!

connectED.mcgraw-hill.com

Chapter

1 Apply Addition and Subtraction Concepts

ESSENTIAL QUESTION

What strategies can I use to add and subtract?

Let's Find Some Animals!

Watch a video!

Watch

Chapter 1 Project

Math Posters

1. Write one of the vocabulary words given by your teacher below.
2. Work with your group to define or draw a picture of different vocabulary words on each group member's page.
3. Glue all of the pages onto poster board.
4. Present your Math Poster to the class.

Name _____

Am I Ready?

Add.

1. ＋ _____ bananas

Subtract.

2. ☒☒☒☒☒☒☒☐☐ $9 - 7 =$ _____

Look at the picture. Write the number sentence.
Add.

3.

_____ ◯ _____ ◯ _____ apples

4. Christy put 6 marbles in a group.
2 marbles rolled away. How many
marbles are left?

_____ marbles

How Did I Do?

Shade the boxes to show the problems you answered correctly.

| 1 | 2 | 3 | 4 |

Name

My Math Words

Review Vocabulary

minus sign (−)   plus sign (+)

Write the word *addition* or *subtraction* in the center. Then use a review vocabulary word to describe your word.

Show as a Symbol

Show as a
Number Sentence

Show an Example
Using a Drawing

Show a Non-Example
Using a Drawing

My Vocabulary Cards

 Vocab

✂

Lesson 1–1

add

$4 + 3 = 7$

Lesson 1–1

addend

$6 + 4 = 10$

Lesson 1–7

count back

$9 - 3 = 6$

Lesson 1–2

count on

$4 + 3 = 7$

Lesson 1–7

difference

$5 - 2 = 3$

Lesson 1–3

doubles

$4 + 4 = 8$

Teacher Directions:
Ideas for Use
- Ask students to arrange cards to show an opposite pair. Have them explain the meaning of their paring.

- Write a tally mark on each card every time you read the word in this chapter or use it in your writing. Try to use at least 10 tally marks for each word card.

Any numbers being added together.

To join together sets to find the total or sum. The opposite of *subtract*.

On a number line, start at the greater addend and move forward by the other addend.

On a number line, start at the greater number and move back by the other number.

Two addends that are the same.

The answer to a subtraction problem.

My Vocabulary Cards

Vocab abc

Processes & Practices

Lesson 1–12

fact family

$6 + 7 = 13$ $13 - 6 = 7$
$7 + 6 = 13$ $13 - 7 = 6$

Lesson 1–11

missing addend

$5 + \boxed{} = 9$

$9 - 5 = \boxed{}$

Lesson 1–3

near doubles

$5 + 6 = 11$

Lesson 1–10

related facts

$3 + 4 = 7$ $4 + 3 = 7$
$7 - 4 = 3$ $7 - 3 = 4$

Lesson 1–7

subtract

$5 - 2 = 3$

Lesson 1–1

sum

$2 + 4 = 6$

Teacher Directions:
More ideas for Use
• Create riddles for each word. Have students guess the word for each riddle.

• Have students draw examples for each card. Have them make drawings that are different from what is shown on each card.

The missing number in a number sentence that makes the number sentence true.

Addition and subtraction sentences that use the same numbers.

Basic facts using the same number. Sometimes called a *fact family*.

Addition facts in which one addend is exactly I more or I less than the other addend.

The answer to an addition problem.

Take away, take apart, separate, or find the difference between two sets. The opposite of *add*.

My Foldable

FOLDABLES® **Follow the steps on the back to make your Foldable.**

$2 + 1$

$6 + 4$

$8 + 7$

$3 + 9$

$7 + 5$

$4 + 0$

___ ◯ ___

___ ◯ ___

1 + 2 =   3

___ ◯ ___ =   ___

___ ◯ ___ =   ___

___ ◯ ___ =   ___

___ ◯ ___ =   ___

___ ◯ ___ =   ___

___ ◯ ___ =   ___

___ ◯ ___ =   ___

Name ....

Addition Properties

ESSENTIAL QUESTION
What strategies can I use
to add and subtract?

*I go bananas
for addition!*

Explore and Explain Watch ▶

I'm Sue!

_____ + _____ = _____

I'm Pat!

_____ + _____ = _____

 Teacher Directions: Draw a picture and write a number sentence to solve. Sue and Pat are zookeepers. Sue fed 5 monkeys and 3 elephants. How many animals did she feed? Pat fed 3 tigers and 5 giraffes. How many animals did he feed?

Processes & Practices

You **add** to find a sum. Each number you add is called an **addend**. The answer is the **sum**.

$3 + 5 =$ ___ sum
↑ ↑
addends

$5 + 3 =$ ___ sum
↑ ↑
addends

Any number plus zero equals that number. This is the Identity Property.

$$\begin{array}{r} 5 \\ + 0 \\ \hline 5 \end{array}$$ ← addends
← sum

$$\begin{array}{r} 0 \\ + 5 \\ \hline 5 \end{array}$$ ← addends
← sum

Find each sum.

1.

   $4 + 3 =$ ___

   $3 + 4 =$ ___

2.

   $6 + 0 =$ ___

   $0 + 6 =$ ___

3.

   $$\begin{array}{r} 2 \\ + 4 \\ \hline \end{array} \qquad \begin{array}{r} 4 \\ + 2 \\ \hline \end{array}$$

4.

   $$\begin{array}{r} 2 \\ + 5 \\ \hline \end{array} \qquad \begin{array}{r} 5 \\ + 2 \\ \hline \end{array}$$

Talk Math Why is the sum the same when you find $3 + 2$ or $2 + 3$?

Name _____

On My Own

Find each sum.

5. 5 + 1 = ____

1 + 5 = ____

6.
```
  4      5
+ 5    + 4
```

7.
```
  0      3
+ 3    + 0
```

8.
```
  6      3
+ 3    + 6
```

9.
```
  7      1
+ 1    + 7
```

10. 6 + 2 = _____

_____ = 2 + 6

11. 8 + 0 = _____

0 + 8 = _____

12. _____ = 4 + 2

2 + 4 = _____

13. 3 + 4 = _____

4 + 3 = _____

14. 0 + 9 = _____

9 + 0 = _____

Problem Solving

15. What two addition facts could you use to find the total number of dots on this domino?

____ + ____ = ____   ____ + ____ = ____

---

16. Manuel's team scores 卌| runs in the first game. They score ||| runs in the second game. Show two ways you can find the total number of runs.

____ + ____ = ____   ____ + ____ = ____

Write Math   Write what you know about the order of the addends in an addition problem.

_____

_____

_____

Name _____

My Homework

Homework Helper Need help? connectED.mcgraw-hill.com

You can add in any order to find a sum.

$$5 \quad 0$$
$$+0 \quad +5$$
$$\overline{5} \quad \overline{5}$$

$$4 + 5 = 9$$
$$5 + 4 = 9$$

Practice

Find each sum.

1. $\begin{array}{r} 2 \\ +4 \\ \hline \end{array}$ $\begin{array}{r} 4 \\ +2 \\ \hline \end{array}$

2. $\begin{array}{r} 4 \\ +1 \\ \hline \end{array}$ $\begin{array}{r} 1 \\ +4 \\ \hline \end{array}$

3. $\begin{array}{r} 9 \\ +0 \\ \hline \end{array}$ $\begin{array}{r} 0 \\ +9 \\ \hline \end{array}$

4. $\begin{array}{r} 3 \\ +4 \\ \hline \end{array}$ $\begin{array}{r} 4 \\ +3 \\ \hline \end{array}$

5. $1 + 7 =$ _____

    $7 + 1 =$ _____

6. _____ $= 0 + 3$

    _____ $= 3 + 0$

Find each sum.

7. 2 + 3 = _____

   3 + 2 = _____

8. 5 + 3 = _____

   3 + 5 = _____

9. The zoo has 4 striped snakes. It has 2 yellow snakes. Write two ways you can find how many snakes the zoo has in all.

Slither!

_____ + _____ = _____   _____ + _____ = _____

10. Kathy counts 4 baby goats and 5 baby sheep in the petting zoo. Write two ways you can find how many babies there are in all?

_____ + _____ = _____   _____ + _____ = _____

Vocabulary Check

Circle the number sentence that shows the vocabulary word.

11. **addends**   3 + 1 = 4   3 + 1 = 4   3 + 1 = 4
                     ↑   ↑   ↑   ↑   ↑

Math at Home Have your child solve the following problem: There are three kittens eating. Two more come over to eat. How many kittens are there in all?

Name _____

Count On to Add

Lesson 2

ESSENTIAL QUESTION
What strategies can I use to add and subtract?

Explore and Explain

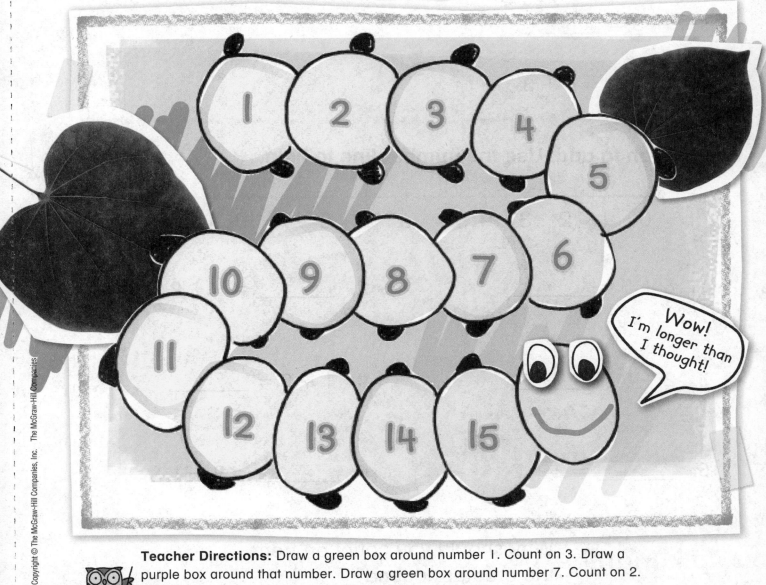

Wow! I'm longer than I thought!

Teacher Directions: Draw a green box around number 1. Count on 3. Draw a purple box around that number. Draw a green box around number 7. Count on 2. Draw a purple box around that number. Draw a green box around number 11. Count on 1. Draw a purple box around that number.

Online Content at **connectED.mcgraw-hill.com**

See and Show

Find 7 + 3. Use a number line. **Count on** to add.
Start with the greater addend.

Start at 7. Count on 3.

$7 + 3 =$ __10__

__10__ $= 7 + 3$

Count on to add. Use the number line to help.

1. $6 + 3 =$ _____

2. _____ $= 4 + 2$

3. $2 + 9 =$ _____

4. $1 + 3 =$ _____

5. _____ $= 2 + 7$

6. _____ $= 9 + 1$

Talk Math   Why should you count on from
the greater addend?

Name _____

On My Own

Count on to add. Use the number line to help.

Helpful Hint
Start with the greater addend.

7. $7 + 1 =$ _____

8. _____ $= 5 + 2$

9. $3 + 8 =$ _____

10. $6 + 1 =$ _____

11. _____ $= 9 + 3$

12. $2 + 8 =$ _____

13.
$$\begin{array}{r} 2 \\ + 7 \\ \hline \end{array}$$

14.
$$\begin{array}{r} 3 \\ + 2 \\ \hline \end{array}$$

15.
$$\begin{array}{r} 1 \\ + 9 \\ \hline \end{array}$$

16.
$$\begin{array}{r} 5 \\ + 1 \\ \hline \end{array}$$

17.
$$\begin{array}{r} 3 \\ + 6 \\ \hline \end{array}$$

18.
$$\begin{array}{r} 1 \\ + 4 \\ \hline \end{array}$$

19.
$$\begin{array}{r} 3 \\ + 1 \\ \hline \end{array}$$

20.
$$\begin{array}{r} 2 \\ + 6 \\ \hline \end{array}$$

21.
$$\begin{array}{r} 8 \\ + 1 \\ \hline \end{array}$$

22. Annie buys 6 eggs at the market.
She has 3 more eggs at home.
How many eggs does she have in all?

Anybody in there?

_____ eggs

23. Sal's cow gave 3 pails of milk in the
morning and 5 pails in the afternoon.
How many pails of milk did Sal's
cow give?

_____ pails

24. Joseph has 4 pigs on his farm. One
pig has 3 piglets. How many pigs
are on the farm?

_____ pigs

Write Math How does counting on help you add?

_____

_____

_____

Name _____

My Homework

Homework Helper **Need help?** connectED.mcgraw-hill.com

Use a number line to find 9 + 3. Start with the greater addend.
Count on to add.

$$9 + 3 = 12$$

Practice

Count on to add. Use the number line above.

1. 4 + 1 = _____

2. 2 + 6 = _____

3. 1 + 8 = _____

4. 3 + 6 = _____

5. 2 + 4 = _____

6. 2 + 8 = _____

7.   3
  + 9
___

8.   4
  + 3
___

9.   1
  + 4
___

Read and then solve the problems.

10. 6 bees are buzzing near a hive.
3 more bees come out of the hive.
How many bees are there now?

_____ bees

11. Cherie has 5 trading cards. She gets
3 more cards. How many cards does
Cherie have now?

I love baseball!

_____ cards

12. There are 8 butterflies on a bush. 3 more
butterflies fly onto the bush. How many
butterflies are on the bush now?

Me too!

_____ butterflies

Vocabulary Check

Draw lines to match.

13. **count on**

$6 + 3 = 9$

14. **addend**

```
 0  1  2  3  4  5  6  7
```

Math at Home Give your child an addition fact. Have him or her count on from the greater number to find the sum.

Doubles and Near Doubles

Lesson 3

ESSENTIAL QUESTION
What strategies can I use to add and subtract?

Explore and Explain

Watch · Tools

I am seeing doubles!

_____ + _____ = _____

_____ + _____ = _____

_____ + _____ = _____

 Teacher Directions: Look at each cube train. Draw the same train beside it. Write the doubles fact shown by each pair of cube trains.

See and Show

Two addends that are the same are called **doubles**.

Use doubles facts to find the sum.

$$6 + 6 = 12$$

addend   addend   sum

We're doubles!

Yeah!

Two addends that are almost a doubles fact are called **near doubles**.

Use near doubles facts to find the sum.

$$6 + 7 = 13$$

addend   addend   sum

Add. Circle the doubles facts.

1. $4 + 4 =$ _____

2. $3 + 4 =$ _____

3. _____ $= 7 + 7$

4. _____ $= 9 + 9$

5. $7 + 8 =$ _____

6. _____ $= 8 + 8$

Talk Math   How can you use doubles and near doubles to remember $5 + 5$? $5 + 6$?

On My Own

Add. Circle the doubles facts.

7. 0 + 0 = _____

8. 7 + 6 = _____

9. 2 + 8 = _____

10. 8 + 9 = _____

11. 3 + 2 = _____

12. 2 + 9 = _____

13.	2	**14.**	7	**15.**	1	**16.**	9
	+ 2		+ 7		+ 1		+ 3

Add. Circle the near doubles facts.

17.	4	**18.**	2	**19.**	7	**20.**	3
	+ 3		+ 4		+ 8		+ 8

21.	2	**22.**	9	**23.**	3	**24.**	4
	+ 7		+ 8		+ 5		+ 5

25. Mr. Bean sells 5 melons to Ed. He sells
the same number of melons to Jose.
How many melons did Mr. Bean sell in all?

I'm hungry!

_____ melons

26. Andrea has 3 dogs. Juanita has the same
number of dogs. How many dogs do
they have in all?

_____ dogs

27. Anthony has 3 lizards. His cousin has one
more lizard than he does. How many lizards
do Anthony and his cousin have in all?

_____ lizards

Write Math   How does knowing doubles facts help
you with solving near doubles facts?

_____

_____

_____

My Homework

Homework Helper Need help? connectED.mcgraw-hill.com

Two addends that are the same are called doubles.

Two addends that are almost a doubles fact are
called near doubles.

Doubles

$4 + 4 = 8$

Near Doubles

$4 + 5 = 9$

Practice

Add. Circle the doubles facts.

1. $5 + 7 =$ _____

2. $2 + 3 =$ _____

3. $5 + 5 =$ _____

4. $8 + 7 =$ _____

5. $\begin{array}{r} 4 \\ + 2 \\ \hline \end{array}$

6. $\begin{array}{r} 4 \\ + 4 \\ \hline \end{array}$

7. $\begin{array}{r} 6 \\ + 7 \\ \hline \end{array}$

8. $\begin{array}{r} 8 \\ + 8 \\ \hline \end{array}$

Add. Circle the near doubles facts.

9. $2 + 9 =$ _____

10. $12 + 3 =$ _____

11. $4 + 5 =$ _____

12. $8 + 8 =$ _____

13. $9 + 8 =$ _____

14. $7 + 6 =$ _____

Solve. Then write the doubles fact that helped you solve the problem.

15. 4 frogs are sitting beside a pond. 5 dragon flies land beside them. How many animals are there altogether?

_____ $+$ _____ $=$ _____ animals

_____ $+$ _____ $=$ _____

Minus one dragon fly!

Vocabulary Check

Draw lines to match.

16. **near doubles** $4 + 4 = 8$

17. **doubles** $4 + 5 = 9$

Math at Home Have your child use objects to make doubles and tell the addition fact.

Make a 10

Lesson 4

ESSENTIAL QUESTION
What strategies can I use to add and subtract?

Explore and Explain Watch Tools

 You can count on us!

$$\underline{\hspace{2cm}} + \underline{\hspace{2cm}} = \underline{\hspace{2cm}}$$

$$\underline{\hspace{2cm}} + \underline{\hspace{2cm}} = \underline{\hspace{2cm}}$$

 Teacher Directions: Place 13 counters in a cup. Empty the cup onto the page. Place red counters in the top ten-frame. Place yellow counters in the bottom ten-frame. Write the addition sentence. Move yellow counters into the top ten-frame to make ten. Write the new addition sentence.

See and Show

You can make a ten to help you add. Find 8 + 4.

First:
Show 8.
Then show 4.

8 + 4

Next:
Take apart 4
to make a 10.

8 + 2 = 10

Last:
Add.

Now you
have 10 + 2.

$10 +$ __2__ $=$ __12__

So, 8 + 4 = __12__.

Use Work Mat 2 and ●●. **Make a 10 to add.**

1.

9 + 5 = _____

2.

7 + 5 = _____

Talk Math   Name all of the facts you know
that have a sum of 10.

Name _____

Helpful Hint
Make a 10, count on,
or find doubles to add.

On My Own

Use Work Mat 2 and . Make a 10 to add.

3. 8 + 6 = _____

4. 7 + 8 = _____

5. 9 + 4 = _____

6. 7 + 7 = _____

7. 4 + 8 = _____

8. 8 + 9 = _____

9. 3 + 9 = _____

10. 7 + 5 = _____

11.   3
         + 7
        ‾‾‾‾

12.   4
         + 9
        ‾‾‾‾

13.   2
         + 9
        ‾‾‾‾

14.   7
         + 8
        ‾‾‾‾

15.   8
         + 4
        ‾‾‾‾

16.   4
         + 6
        ‾‾‾‾

Problem Solving

17. Ms. Ling's class sees 9 tigers at the zoo.
Ms. John's class sees 8 tigers.
How many tigers did they see in all?

_____ tigers

18. There are 6 pigs at the farm. There are 7 sheep at the farm. How many pigs and sheep are there in all?

Did I just see a tiger?

_____ pigs and sheep

19. 5 goldfish are in the tank. 7 more goldfish are added to the tank. How many goldfish are in the tank now?

_____ goldfish

 How could you use making a ten when adding?

_____

_____

_____

Name _____

My Homework

Homework Helper Need help? connectED.mcgraw-hill.com

Make a ten to help you add.

8 + 6

Think: 8 + 2 = 10
Add: 10 + 4 = 14.
So, 8 + 6 = 14.

Practice

Make a ten to add.

1.

6 + 5 = _____

2.

8 + 4 = _____

Make a ten to add.

3. $7 + 4 =$ _____

4. $4 + 8 =$ _____

5. $9 + 7 =$ _____

6. $7 + 6 =$ _____

7. $3 + 9 =$ _____

8. $7 + 5 =$ _____

9.
$$\begin{array}{r} 8 \\ + 8 \\ \hline \end{array}$$

10.
$$\begin{array}{r} 9 \\ + 4 \\ \hline \end{array}$$

11.
$$\begin{array}{r} 8 \\ + 5 \\ \hline \end{array}$$

Raul is unstoppable!

12. Raul wins 8 chess matches on Saturday and 5 on Sunday. How many matches did he win?

_____ matches

Test Practice

13. Which addition sentence can help you find the sum for $8 + 7$?

$5 + 7 = 12$ ○

$10 + 5 = 15$ ○

$10 + 8 = 18$ ○

$10 + 7 = 17$ ○

Math at Home Have your child tell you how to find $8 + 7$ by making a 10.

Name

Add Three Numbers

Lesson 5

ESSENTIAL QUESTION
What strategies can I use to add and subtract?

Explore and Explain

Watch ▶

Tools

_____ animals

 Teacher Directions: Have students use counters to show the following: There are 4 fish, 6 frogs, and 2 snakes in the pond. Trace the counters and write how many in all.

See and Show

Helpful Hint
Look for facts you know.

You can group addends in different ways. The sum will stay the same.

$4 + 6 + 4$

$2 + 3 + 4$

$3 + 1 + 7$

$6 + 8 =$ __14__ $5 + 4 =$ __9__ $1 + 10 =$ __11__

Find each sum. Circle the numbers you add first.
Write that sum in the box.

1. $2 + 4 + 2 =$ _____

2. $9 + 1 + 2 =$ _____

3. $7 + 4 + 7 =$ _____

4. $2 + 7 + 8 =$ _____

5. $3 + 7 + 4 =$ _____

6. $5 + 3 + 5 =$ _____

Talk Math How do you decide which two numbers to add first when adding three digits?

On My Own

Find each sum. Circle the numbers you add first. Write that sum in the box.

7. $4 + 3 + 4 =$ _____

8. $4 + 3 + 7 =$ _____

9. $2 + 8 + 3 =$ _____

10. $6 + 1 + 6 =$ _____

Find each sum.

11. $5 + 5 + 5 =$ _____

12. $6 + 6 + 3 =$ _____

13. $7 + 4 + 7 =$ _____

14. $9 + 8 + 1 =$ _____

15. $1 + 7 + 3 =$ _____

16. $5 + 7 + 5 =$ _____

17.
$$\begin{array}{r} 3 \\ 5 \\ + 7 \\ \hline \end{array}$$

18.
$$\begin{array}{r} 7 \\ 7 \\ + 1 \\ \hline \end{array}$$

19.
$$\begin{array}{r} 9 \\ 8 \\ + 1 \\ \hline \end{array}$$

Problem Solving

Processes & Practices

20. The zoo has 5 black bears, 5 brown bears and 2 polar bears. How many bears are at the zoo?

I can't wait to see the answer!

_____ bears

21. Lisa saw 4 sheep and 4 goats at the petting zoo. She also saw 3 piglets. How many animals did Lisa see at the petting zoo?

_____ animals

Write the doubles fact that helped you solve the problem.

_____ + _____ = _____

Explain how you can group addends in any way to help you solve addition problems that have more than 2 addends.

_____

_____

_____

Name _____

My Homework

Homework Helper **Need help?** connectED.mcgraw-hill.com

Look for facts you know to help you add three numbers.

6 + 5 + 4

| 10 |

10 + 5 = 15

4 + 5 + 4

| 8 |

8 + 5 = 13

3 + 1 + 4

| 4 |

4 + 4 = 8

Practice

Find each sum.

1. 6 + 6 + 4 = _____

2. 6 + 2 + 8 = _____

3. 3 + 3 + 9 = _____

4. 7 + 4 + 3 = _____

5.
   1
   9
+ 4

6.
   7
   6
+ 6

7.
   8
   4
+ 2

Find each sum.

8.
```
    7
    3
  + 5
```

9.
```
    4
    2
  + 6
```

10.
```
    9
    8
  + 1
```

11. The zoo has 3 giraffes, 4 African elephants, and 3 Asian elephants. How many elephants and giraffes are at the zoo?

I've got lots of elephant and giraffe friends!

_____ giraffes and elephants

12. There are 5 tigers, 4 leopards, and 6 lions at the zoo. How many big cats are at the zoo?

_____ big cats

Test Practice

13. Which addition sentence could help solve this problem?

$$3 + 5 + 7 = \underline{\hspace{1cm}}$$

$2 + 3 = 5$ ○ $5 + 5 = 10$ ○ $3 + 7 = 10$ ○ $5 + 2 = 7$ ○

Math at Home Have your child show you how to add 7 + 7 + 1.

Problem Solving

STRATEGY: Write a Number Sentence

Lesson 6

ESSENTIAL QUESTION
What strategies can I use
to add and subtract?

Four friends were eating a snack.
Two more friends joined them. Each person
ate a banana. How many bananas did they eat?

Don't slip!

1 Understand Underline what you know.
Circle what you need to find.

2 Plan How will I solve the problem?

3 Solve Write a number sentence.

$$4 + 2 = 6$$

_____ bananas

Too late!

4 Check Is my answer reasonable? Explain.

Practice the Strategy

Mrs. Brown's class had a pet parade.
3 students each brought a brown rabbit.
2 students each brought a white rabbit.
I student brought a black rabbit. How many
rabbits were in the parade?

1 Understand
Underline what you know. Circle what you need to find.

2 Plan
How will I solve the problem?

3 Solve
I will...

4 Check
Is my answer reasonable? Explain.

Processes & Practices

Apply the Strategy

Write a number sentence to solve.

1. There are 3 rabbits at a farm. The farmer buys 9 more. How many rabbits are there now?

_____ ◯ _____ ◯ _____   _____ rabbits

2. Chung always picks 5 flowers at home after school. How many flowers will he have after 3 days?

_____ ◯ _____ ◯ _____ ◯ _____   _____ flowers

3. Mrs. Lewis has 4 books for her reading group. She finds 2 more books. How many books does she have now?

_____ ◯ _____ ◯ _____   _____ books

Review the Strategies

Choose a strategy
- Write a number sentence.
- Act it out.
- Draw a diagram.

4. Steve, Elena, and Ian are playing with toy planes. Steve gave 8 planes to Ian and 8 to Elena. He has 4 planes left. How many planes did Steve start with?

_____ planes

5. Arielle has 4 markers. Julia has 3 more markers than Arielle. How many markers do they have in all?

_____ markers

6. The library has 4 books about parrots and 3 books about swans. There is 1 less book about blue jays than swans. How many books about birds are there in all?

Parrot books are my favorite!

_____ books

Name ....

My Homework

Kim's mom is making blueberry pancakes.
She makes 4 pancakes in the first batch.
In the second batch she makes 5 more.
How many pancakes does Kim's mom make?

1 Understand Underline what you know.
Circle what you need to find.

2 Plan How will I solve the problem?

3 Solve I will write a number sentence.

$4 + 5 = 9$

Kim's mom made 9 pancakes.

4 Check Is my answer reasonable?

Practice

Underline what you know. Circle what you need to find. Write a number sentence to solve.

1. Kit, Ruth, and Dean each eat 2 slices of pizza. How much pizza do they eat in all?

   _____ + _____ + _____ = _____ pizza slices

I have 9 sweaters in me!

2. Aki packs 2 suitcases with 9 sweaters in each. How many sweaters did he pack in all?

   _____ + _____ = _____ sweaters

3. Kylee is making a pattern with 6 circles and 3 squares. How many shapes does she use in all?

   _____ + _____ = _____ shapes

Me, too! It must be cold where we're going!

Test Practice

4. Which addition sentence is not correct?

   $4 + 4 + 2 = 10$ ○ $4 + 3 + 3 = 10$ ○

   $4 + 2 + 2 = 8$ ○ $4 + 2 + 4 = 8$ ○

Math at Home Have your child write a number sentence to solve this problem. There are two kittens eating. Three more come over to eat. How many kittens are there in all?

Name ....

Check My Progress

Vocabulary Check

Draw lines to match.

1. **add**   Both addends are the same.

2. **doubles** $3 + 4 = 7$

3 4 5 6 7

3. **count on**   The numbers you add in an addition number sentence.

4. **addends** $6 + 8 = 14$

Concept Check

Find each sum.

5. $7 + 0 =$ _____   6. $3 + 6 =$ _____   7. $2 + 5 =$ _____

   $0 + 7 =$ _____ $6 + 3 =$ _____ $5 + 2 =$ _____

Add. Circle the doubles facts.

8. $\begin{array}{r} 0 \\ + 3 \\ \hline \end{array}$ 9. $\begin{array}{r} 1 \\ + 1 \\ \hline \end{array}$ 10. $\begin{array}{r} 3 \\ + 4 \\ \hline \end{array}$ 11. $\begin{array}{r} 9 \\ + 9 \\ \hline \end{array}$

Make a ten to add.

12. $\begin{array}{r} 4 \\ + 8 \\ \hline \end{array}$ 13. $\begin{array}{r} 9 \\ + 5 \\ \hline \end{array}$ 14. $\begin{array}{r} 7 \\ + 6 \\ \hline \end{array}$ 15. $\begin{array}{r} 6 \\ + 8 \\ \hline \end{array}$

Find each sum.

16. $\begin{array}{r} 1 \\ 4 \\ + 6 \\ \hline \end{array}$ 17. $\begin{array}{r} 5 \\ 2 \\ + 8 \\ \hline \end{array}$ 18. $\begin{array}{r} 7 \\ 1 \\ + 2 \\ \hline \end{array}$ 19. $\begin{array}{r} 8 \\ 8 \\ + 2 \\ \hline \end{array}$

20. Lou saw 6 penguins at the zoo.
He also saw 3 flamingos and 2 pelicans.
How many birds did he see in all?

_____ birds

Test Practice

21. Ali's dog had 8 puppies. Luca's dog had the same
number of puppies. How many puppies are there in all?

8   16   18   88
○   ○   ○   ○

Name _____

Count Back to Subtract

Lesson 7

ESSENTIAL QUESTION
What strategies can I use to add and subtract?

Explore and Explain

Watch ▶ Tools

There are **SO** many to choose from!

1 2 3 4 5

6 7 8 9 10

11 12 13 14 15

16 17 18 19 20

Teacher Directions: Draw a blue box around number 9. Count back 3. Circle that number red. Draw a green box around number 19. Count back 2. Circle that number yellow. Draw a purple box around 14. Count back 1. Circle that number orange.

See and Show

Find 10 − 3. Use a number line. **Count back** to **subtract**. The answer is the **difference**.

Start with the greater number.

$$10 - 3 = \underline{\quad 7 \quad}$$
difference

$$\begin{array}{r} 10 \\ -\ 3 \\ \hline 7 \end{array}$$
difference

No flowers on this page! I want to go <u>back</u>.

Count back to subtract. Use the number line.

1. 7 − 3 = _____

2. 8 − 1 = _____

3. 9 − 2 = _____

4. 10 − 2 = _____

5. 8 − 2 = _____

6. _____ = 7 − 1

Talk Math   Explain how you count back on a number line to subtract.

Name

_____

On My Own

Count back to subtract. Use the number line.

7. $3 - 2 =$ _____

8. _____ $= 12 - 3$

9. _____ $= 8 - 3$

10. $11 - 3 =$ _____

11. $9 - 3 =$ _____

12. $11 - 2 =$ _____

13.
$$\begin{array}{r} 9 \\ -\ 1 \\ \hline \end{array}$$

14.
$$\begin{array}{r} 5 \\ -\ 3 \\ \hline \end{array}$$

15.
$$\begin{array}{r} 10 \\ -\ 1 \\ \hline \end{array}$$

16.
$$\begin{array}{r} 4 \\ -\ 2 \\ \hline \end{array}$$

17.
$$\begin{array}{r} 12 \\ -\ 2 \\ \hline \end{array}$$

18.
$$\begin{array}{r} 4 \\ -\ 3 \\ \hline \end{array}$$

19.
$$\begin{array}{r} 8 \\ -\ 1 \\ \hline \end{array}$$

20.
$$\begin{array}{r} 7 \\ -\ 2 \\ \hline \end{array}$$

21.
$$\begin{array}{r} 5 \\ -\ 2 \\ \hline \end{array}$$

Problem Solving

22. Tanya has 12 books. She gives 3 away.
How many books does she have now?

_____ books

23. Hank washes 9 windows at the pet store. He
washes 2 windows in the morning. How many
windows does he wash in the afternoon?

squawk!

_____ windows

24. The animal shelter had 8 birds. 3 of the birds
were adopted. How many birds were left at the
animal shelter?

_____ birds

 Explain how counting back can
help you subtract.

_____

_____

_____

Name _____

My Homework

Homework Helper **Need help?** connectED.mcgraw-hill.com

Use the number line. Start at 9, count back to subtract.

$$9 - 3 = 6$$

Practice

Count back to subtract. Use the number line above.

1. $12 - 4 =$ _____

2. $11 - 3 =$ _____

3. $8 - 3 =$ _____

4. $6 - 2 =$ _____

5. $6 - 3 =$ _____

6. $10 - 2 =$ _____

7. $\begin{array}{r} 12 \\ -\ 3 \\ \hline \end{array}$

8. $\begin{array}{r} 11 \\ -\ 1 \\ \hline \end{array}$

9. $\begin{array}{r} 8 \\ -\ 2 \\ \hline \end{array}$

10. A paper clip holder has 12 clips.
Alex uses 3 paper clips.
How many paper clips are left?

_____ paper clips

11. Marty buys 11 erasers. He uses 1 eraser.
How many erasers does Marty have left?

_____ erasers

12. Jenny had 9 pencils in her desk. She let
her friends borrow 2 of them. How many
pencils does she have left?

Pick me!

Pick me instead!

_____ pencils

Vocabulary Check

Complete the sentence.

| count back | sum | difference |

13. You can _____ to subtract.

14. The answer to a subtraction problem is
called the _____.

 Math at Home Have your child count back to subtract. Say a number between 3 and 12.
Have you child subtract 1, 2, or 3.

Subtract All and Subtract Zero

Lesson 8

ESSENTIAL QUESTION
What strategies can I use to add and subtract?

Bubbles tickle my nose.

Explore and Explain
Watch | Tools

I'm drenched!

_____ − _____ = _____ dogs

Teacher Directions: Draw a picture and write a number sentence to solve the problem. Mary helped wash dogs at the pet store. There were 7 dogs to be washed. Mary washed 7 of them. How many dogs were left to be washed?

Online Content at connectED.mcgraw-hill.com

See and Show

You can subtract to find the difference.
Find how many frogs are left.

Subtract all.	Subtract zero.

$5 - 5 = \underline{0}$

$5 - 0 = \underline{5}$

$\underline{0}$ frogs are left.

$\underline{5}$ frogs are left.

Subtract.

1. $13 - 13 = \underline{\hspace{1.5cm}}$

   $13 - 0 = \underline{\hspace{1.5cm}}$

2. $\underline{\hspace{1.5cm}} = 7 - 0$

   $7 - 7 = \underline{\hspace{1.5cm}}$

3. $6 - 6 = \underline{\hspace{1.5cm}}$

   $\underline{\hspace{1.5cm}} = 6 - 0$

4. $8 - 0 = \underline{\hspace{1.5cm}}$

   $8 - 8 = \underline{\hspace{1.5cm}}$

Talk Math Explain how you know $8 - 8 = 0$ and $8 - 0 = 8$.

On My Own

Subtract. Circle the problem if the difference is zero.

5. $9 - 3 =$ _____

6. _____ $= 8 - 1$

7. _____ $= 4 - 4$

8. $7 - 3 =$ _____

9.
$$\begin{array}{r} 8 \\ -\ 8 \\ \hline \end{array}$$

10.
$$\begin{array}{r} 8 \\ -\ 2 \\ \hline \end{array}$$

11.
$$\begin{array}{r} 6 \\ -\ 0 \\ \hline \end{array}$$

12.
$$\begin{array}{r} 5 \\ -\ 5 \\ \hline \end{array}$$

13.
$$\begin{array}{r} 3 \\ -\ 2 \\ \hline \end{array}$$

14.
$$\begin{array}{r} 16 \\ -\ 16 \\ \hline \end{array}$$

15.
$$\begin{array}{r} 7 \\ -\ 2 \\ \hline \end{array}$$

16.
$$\begin{array}{r} 4 \\ -\ 0 \\ \hline \end{array}$$

17.
$$\begin{array}{r} 9 \\ -\ 0 \\ \hline \end{array}$$

18.
$$\begin{array}{r} 8 \\ -\ 0 \\ \hline \end{array}$$

19.
$$\begin{array}{r} 9 \\ -\ 9 \\ \hline \end{array}$$

20.
$$\begin{array}{r} 12 \\ -\ 2 \\ \hline \end{array}$$

21.
$$\begin{array}{r} 11 \\ -\ 6 \\ \hline \end{array}$$

22.
$$\begin{array}{r} 3 \\ -\ 3 \\ \hline \end{array}$$

23.
$$\begin{array}{r} 7 \\ -\ 1 \\ \hline \end{array}$$

24.
$$\begin{array}{r} 6 \\ -\ 6 \\ \hline \end{array}$$

Problem Solving

25. 13 bees buzz near a flower. None fly away. How many bees are near the flower?

_____ bees

26. 4 sparrows are in the nest. They all fly away. How many sparrows are still in the nest?

Where did everybody go?

_____ sparrows

27. Todd had 12 beads. He lost some of them. Now he has 7 beads. How many beads did Todd lose?

_____ beads

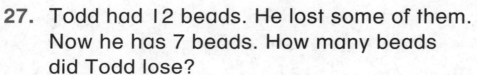
Write Math   In a subtraction number sentence, which number is the difference?

_____

_____

_____

Name

My Homework

Homework Helper **Need help?** connectED.mcgraw-hill.com

You can subtract all.

$6 - 6 = 0$

You can subtract zero.

$6 - 0 = 6$

Practice

Subtract.

1.
$$\begin{array}{r} 9 \\ -\ 0 \\ \hline \end{array}$$

2.
$$\begin{array}{r} 15 \\ -\ 15 \\ \hline \end{array}$$

3.
$$\begin{array}{r} 1 \\ -\ 1 \\ \hline \end{array}$$

4.
$$\begin{array}{r} 7 \\ -\ 0 \\ \hline \end{array}$$

5.
$$\begin{array}{r} 12 \\ -\ 12 \\ \hline \end{array}$$

6.
$$\begin{array}{r} 4 \\ -\ 4 \\ \hline \end{array}$$

7.
$$\begin{array}{r} 8 \\ -\ 0 \\ \hline \end{array}$$

8.
$$\begin{array}{r} 0 \\ -\ 0 \\ \hline \end{array}$$

9. $3 - 0 =$ _____

10. $17 - 17 =$ _____

11. $3 - 3 =$ _____

12. $2 - 0 =$ _____

13. 10 children play ball. After they finish, all 10 go back to class. How many children keep playing ball?

_____ children

14. 8 girls take a walk. When they reach the park, they all keep walking. How many girls are still taking a walk?

_____ girls

15. 18 children are at the store. All of them check out and go outside. How many are still shopping?

_____ children

Test Practice

16. Find the difference.

$19 - 19 =$

0	18	19	38
○	○	○	○

Math at Home Have your child use small objects to show 5 − 5 and 5 − 0.

Use Doubles to Subtract

Lesson 9

ESSENTIAL QUESTION
What strategies can I use to add and subtract?

Explore and Explain

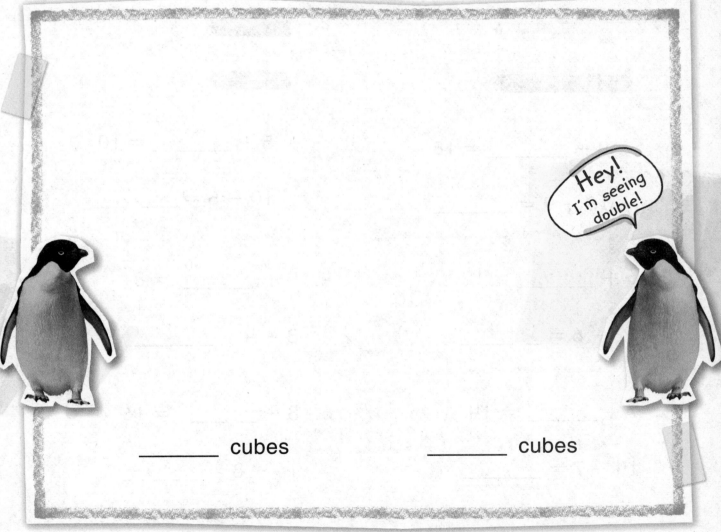

Hey! I'm seeing double!

_____ cubes   _____ cubes

 Teacher Directions: Place a 6 cube train at the top of the page. Trace it. Separate the train into two equal parts. Draw the parts below the 6 cube train. Write how many cubes are in each part.

See and Show

You can use doubles facts to help you subtract.
Find 16 − 8.

If you know 8 + 8 = 16,

then 16 − 8 = _____.

Use cubes and doubles facts to subtract.

1.

   9 + _____ = 18

   18 − 9 = _____

2.

   5 + _____ = 10

   10 − 5 = _____

3. 6 + _____ = 12

   12 − 6 = _____

4. 4 + _____ = 8

   8 − 4 = _____

5. 7 + _____ = 14

   14 − 7 = _____

6. 8 + _____ = 16

   16 − 8 = _____

Talk Math Explain how you can use a doubles fact
to subtract.

Name _____

Helpful Hint
I Know 5 + 5 = 10,
so 10 − 5 = 5.

On My Own

Subtract. Circle the doubles facts.

7. 9 − 3 = _____

8. 11 − 9 = _____

9. 6 − _____ = 3

10. 7 − 3 = _____

11. 18 − 9 = _____

12. 15 − 7 = _____

13.
```
    16
 -  □
 ─────
     8
```

14.
```
     7
  −  0
 ─────
```

15.
```
    10
 -  □
 ─────
     5
```

16.
```
     9
  −  8
 ─────
```

17.
```
     6
  −  2
 ─────
```

18.
```
     8
  −  3
 ─────
```

19.
```
    14
 -  □
 ─────
     7
```

20.
```
     7
  −  6
 ─────
```

21.
```
     8
 -  □
 ─────
     4
```

22. Fran and her grandmother pick 16 pumpkins. They use 8 pumpkins for pie. How many pumpkins are left?

_____ pumpkins

23. Delia bakes 18 cherry pies. She sells 9 pies at a farmers' market. How many pies does Delia have left?

_____ pies

24. 14 children were playing hide and go seek. 7 children were found. How many children are still hiding?

_____ children

Write Math Explain how you can use doubles facts to help you subtract.

_____

_____

_____

Name _____

My Homework

Homework Helper **Need help?** connectED.mcgraw-hill.com

Use doubles facts to subtract. Find $12 - 6$.

If you know $6 + 6 = 12$,

then you also know $12 - 6 = 6$.

Practice

Subtract. Circle the doubles facts.

1. $18 - 9 =$ _____

2. $6 - 2 =$ _____

3. $8 - 2 =$ _____

4. $5 - 3 =$ _____

5. $16 - 8 =$ _____

6. $8 - 3 =$ _____

7. $6 - 3 =$ _____

8. $8 - 8 =$ _____

9. $\begin{array}{r} 14 \\ -\ 7 \\ \hline \end{array}$

10. $\begin{array}{r} 9 \\ -\ 1 \\ \hline \end{array}$

11. $\begin{array}{r} 10 \\ -\ 4 \\ \hline \end{array}$

Subtract. Circle the doubles facts.

12. $14 - 6 =$ _____

13. $18 - 8 =$ _____

14. $18 - 9 =$ _____

15. $15 - 7 =$ _____

16. Anita checks out 14 library books. She reads 7 of the books. How many books does she still have to read?

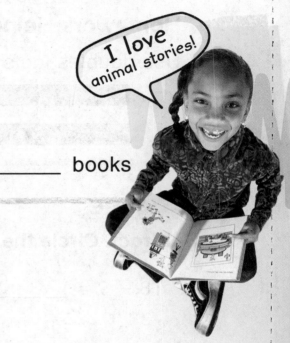

I love animal stories!

_____ books

17. Lindsay has 16 cupcakes to hand out. She has given 8 to the girls in her class. How many does she have left?

_____ cupcakes

Test Practice

18. Mark the number sentence that uses doubles.

$$4 - 4 = 0 \qquad 8 - 4 = 4 \qquad 8 - 3 = 5 \qquad 8 - 8 = 0$$

◯ ◯ ◯ ◯

 Math at Home Call out doubles facts. Have your child name the subtraction problem for each double.

Name _____

Vocabulary Check

Complete each sentence.

count back   subtract   difference

1. You can _____ by taking a number

   away from a greater number.

2. The number you have left after you subtract

   is called the _____.

3. You can _____ on a number line

   to help you subtract.

Concept Check

Count back to subtract. Use the number line to help.

0 1 2 3 4 5 6 7 8 9 10 11 12 13 14 15

4. 13 − 3 = _____   5. 15 − 2 = _____

6. 7 − 3 = _____   7. 14 − 3 = _____

8. 10 − 1 = _____   9. 12 − 8 = _____

Subtract.

10. 6
    − 0

11. 8
    − 8

12. 9
    − 0

13. 4
    − 4

Use doubles or near doubles facts to subtract.

14. 16
    − 8

15. 7
    − 3

16. 10
    − 4

17. 18
    − 9

18. 14
    − 7

19. 8
    − 5

20. 6
    − 3

21. 9
    − 4

---

22. Laine counts 18 fireflies. She counts
    9 fireflies in the front yard. How many
    fireflies does Laine count in the backyard?

    _____ fireflies

Test Practice

23. Ryder rode his bike around the block 16 times. Violet
    rode half as many times. How many times did Violet ride
    her bike around the block?

    16   10   8   6
    ○   ○   ○   ○

Relate Addition and Subtraction

Lesson 10

ESSENTIAL QUESTION
What strategies can I use to add and subtract?

Explore and Explain

 Watch   Tools

Part	Part
Whole	

____ + ____ = ____

____ − ____ = ____

Did those counters move?

 Teacher Directions: Place 4 red counters in one part. Place 3 yellow counters in another part. Move the counters to the whole to show the total. Write the addition sentence. Move the yellow counters back to the part. Write the subtraction sentence shown.

Online Content at ⌐ **connectED.mcgraw-hill.com**

See and Show

You can use addition facts to subtract.
Related facts have the same three numbers.

5 + 4 = _9_

Helpful Hint
Addition and subtraction are opposite or inverse operations.

9 − 4 = _5_   9 − 5 = _4_

Use the addition facts to subtract.

1.

    6 + 7 = _____

    13 − 6 = _____

    13 − 7 = _____

2.

    5 + 7 = _____

    12 − 7 = _____

    12 − 5 = _____

3.   5 + 6 = _____

       11 − 6 = _____

       11 − 5 = _____

4.   7 + 4 = _____

       11 − 4 = _____

       11 − 7 = _____

Talk Math Explain how addition and subtraction are related.

Name _____

On My Own

Use addition facts to subtract.

5. $3 + 9 =$ _____

$12 - 9 =$ _____

$12 - 3 =$ _____

6. $4 + 7 =$ _____

$11 - 4 =$ _____

$11 - 7 =$ _____

7.
$$\begin{array}{cc} 8 & 17 \\ +\ 9 & -\ 8 \\ \hline \end{array}$$

8.
$$\begin{array}{cc} 9 & 15 \\ +\ 6 & -\ 9 \\ \hline \end{array}$$

9.
$$\begin{array}{cc} 9 & 14 \\ +\ 5 & -\ 5 \\ \hline \end{array}$$

10.
$$\begin{array}{cc} 8 & 12 \\ +\ 4 & -\ 8 \\ \hline \end{array}$$

11.
$$\begin{array}{cc} 8 & 15 \\ +\ 7 & -\ 7 \\ \hline \end{array}$$

12.
$$\begin{array}{cc} 8 & 13 \\ +\ 5 & -\ 8 \\ \hline \end{array}$$

13. $8 + 9 =$ ____

$17 - 8 =$ ____

14. $6 + 8 =$ ____

$14 - 6 =$ ____

15. $6 + 7 =$ ____

$13 - 6 =$ ____

16. $3 + 8 =$ ____

$11 - 8 =$ ____

17. $8 + 8 =$ ____

$16 - 8 =$ ____

18. $5 + 5 =$ ____

$10 - 5 =$ ____

19. The tennis team has 16 players.
8 players leave the team. How many
players are still on the team?

_____ players

Write a related fact. _____ ◯ _____ ◯ _____

20. 10 turtles are on a log. 2 turtles jump
in the water and swim away. How
many turtles are left on the log?

_____ turtles

Write a related fact. _____ ◯ _____ ◯ _____

I'm lucky I
found this log!

21. 9 birds are flying. 3 birds fly away.
How many birds are left flying?

_____ birds

Write a related fact. _____ ◯ _____ ◯ _____

Write Math   Write an example of related facts.
Use the numbers 4, 6, and 10.

_____

_____

_____

Name _____

My Homework

Homework Helper Need help? connectED.mcgraw-hill.com

You can use addition facts to subtract.
If you know $6 + 3 = 9$, then you also know that
$9 - 3 = 6$ and $9 - 6 = 3$.

$6 + 3 = 9$

$9 - 3 = 6$ $9 - 6 = 3$

Practice

Use addition facts to subtract.

1. $7 + 5 =$ _____

   $12 - 5 =$ _____

2. $6 + 9 =$ _____

   $15 - 9 =$ _____

3. $8 + 5 =$ _____

   $13 - 5 =$ _____

4. $4 + 7 =$ _____

   $11 - 7 =$ _____

Use addition facts to subtract.

5.
```
  7     10
+ 3    − 3
```

6.
```
  4      4
+ 0    − 0
```

7.
```
  2     10
+ 8    − 2
```

8. Megan sent 12 postcards last week. This week she sent 4 postcards. How many more postcards did she send last week?

_____ postcards

I hope Megan writes to me!

Vocabulary Check

9. **Using 2, 4, and 6, write an example of related facts.**

Math at Home Ask your child to show you an addition sentence with spoons and relate it to subtraction.

74   Chapter 1 • Lesson 10

Name

Missing Addends

Lesson 11

ESSENTIAL QUESTION
What strategies can I use
to add and subtract?

Explore and Explain

$4 + \boxed{} = 9$

$2 + \boxed{} = 8$

$6 + \boxed{} = 11$

$3 + \boxed{} = 12$

$4 + \boxed{} = 8$

$5 + \boxed{} = 10$

If I were a domino,
I'd be worth 4!

 Teacher Directions: How many dots do you see? How many dots should there be in all? Draw more dots on the blank side of each domino until you have that number of dots in all. Complete the number sentence.

Online Content at ✐ **connectED.mcgraw-hill.com**

Chapter 1 • Lesson 11   75

See and Show

You can use a related fact to help you find a **missing addend**.

$9 + \boxed{} = 15$

Think $15 - 9 = 6$.

$9 + \boxed{6} = 15$

6 is the missing addend.

Find the missing addend. Draw that many dots on the domino.

1.

$8 + \boxed{} = 14$

Think: $14 - 8 = \boxed{}$

2.

$5 + \boxed{} = 11$

Think: $11 - 5 = \boxed{}$

3.

$\boxed{} + 4 = 8$

Think: $8 - 4 = \boxed{}$

4.

$7 + \boxed{} = 15$

Think: $15 - 7 = \boxed{}$

Talk Math How do you find the missing addend in $5 + \boxed{} = 13$?

On My Own

Find the missing addend.

5. $9 + \boxed{} = 12$

$12 - 9 = \boxed{}$

6. $7 + \boxed{} = 8$

$8 - 7 = \boxed{}$

7. $\boxed{} + 6 = 12$

$12 - 6 = \boxed{}$

8. $\boxed{} + 2 = 7$

$7 - 2 = \boxed{}$

9. $7 + \boxed{} = 16$

$16 - 7 = \boxed{}$

10. $\boxed{} + 7 = 14$

$14 - 7 = \boxed{}$

11. $9 + \boxed{} = 9$

$9 - 9 = \boxed{}$

12. $6 + \boxed{} = 13$

$13 - 6 = \boxed{}$

13. $4 + \boxed{} = 12$

$12 - 4 = \boxed{}$

14. $5 + \boxed{} = 12$

$12 - 5 = \boxed{}$

Problem Solving

15. There were 14 seals on a rock. Some seals jumped into the water. 9 seals were left on the rock. How many seals jumped into the water?

_____ seals

16. Anna buys 7 plants. She wants 12 plants. How many more plants does Anna need?

_____ plants

17. J.J. wants 16 fish in his aquarium. He has 6 fish now. How many more fish does J.J. need?

_____ fish

HOT Problem Julie found a missing addend like this:

$$7 + \boxed{} = 9 \qquad 9 + 7 = 16$$

Tell what she did wrong. Make it right.

_____

_____

_____

Name _____

My Homework

Homework Helper **Need help?** connectED.mcgraw-hill.com

Use related facts to find missing addends.

$$7 + \boxed{8} = 15 \qquad 15 - 7 = \boxed{8}$$

Practice

Find the missing addend.

1. $8 + \boxed{} = 12$

   $12 - 8 = \boxed{}$

2. $\boxed{} + 6 = 15$

   $15 - 6 = \boxed{}$

3. $8 + \boxed{} = 13$

   $13 - 8 = \boxed{}$

4. $\boxed{} + 7 = 11$

   $11 - 7 = \boxed{}$

5. $8 + \boxed{} = 17$

   $17 - 8 = \boxed{}$

6. $9 + \boxed{} = 12$

   $12 - 9 = \boxed{}$

7. David and his friends are flying 9 kites. Some kites get stuck in a tree. 7 kites are still flying. How many kites are in the tree?

I'm really good at playing catch!

$7 + \boxed{} = 9$ _____ kites

8. The Girl Scouts put some boats in the pond. They leave 9 boats on land. There are 15 boats altogether. How many boats did the scouts put into the pond?

$9 + \boxed{} = 15$ _____ boats

Ouch!

9. 7 boys were riding their bikes. Some of the boys went home. 3 boys are still riding their bikes. How many boys went home?

$3 + \boxed{} = 7$ _____ boys

Vocabulary Check

Complete the sentence.

missing addend   sum

10. When a number sentence only has one addend, you must find the _____ to solve it.

 Math at Home Ask your child to tell you the subtraction fact that will help him or her find the missing addend in $7 + \boxed{} = 15$.

Fact Families

Lesson 12

ESSENTIAL QUESTION
What strategies can I use to add and subtract?

Explore and Explain

I think I see my cousin!

_____ + _____ = _____

_____ + _____ = _____

_____ − _____ = _____

_____ − _____ = _____

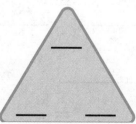

_____ + _____ = _____

_____ + _____ = _____

_____ − _____ = _____

_____ − _____ = _____

 Teacher Directions: Roll a red number cube two times to find two addends. Write the numbers at the bottom of the triangle. Write the sum at the top of the triangle. Use the numbers to write related number sentences. Repeat for the other triangle.

See and Show

A **fact family** is a set of related facts using the same three numbers. The numbers in this fact family are 12, 9, and 3.

$9 + 3 =$ __12__

$3 + 9 =$ __12__

$12 - 9 =$ __3__

$12 - 3 =$ __9__

Helpful Hint
Use related facts to complete fact families.

Complete each fact family.

1.

16
7   9

$9 + 7 =$ _____

$7 + 9 =$ _____

$16 - 9 =$ _____

$16 - 7 =$ _____

2.

17
9   8

$9 + 8 =$ _____

____ $+$ ____ $=$ _____

$17 - 9 =$ _____

____ $-$ ____ $=$ _____

Talk Math   What are the related facts in the fact family 9, 9, 18? Explain.

Name _____

Helpful Hint
Each fact in a fact family uses the same three numbers.

On My Own

Complete each fact family.

3.

___ + ___ = ___

___ + ___ = ___

___ − ___ = ___

___ − ___ = ___

4.

___ + ___ = ___

___ + ___ = ___

___ − ___ = ___

___ − ___ = ___

5.

___ + ___ = ___

___ + ___ = ___

___ − ___ = ___

___ − ___ = ___

6.

___ + ___ = ___

___ + ___ = ___

___ − ___ = ___

___ − ___ = ___

Solve.

7. Linda has 7 fish. She has 5 crabs. How many animals does Linda have?

_____ animals

I used to be shy, but I've come out of my shell!

8. There are 8 apples in one basket. The second basket has one more than the first. How many apples are there in all?

_____ apples

9. 11 children were on the playground. 3 more children came to the playground. How many children are on the playground now?

_____ children

HOT Problem Make a fact family. Use at least one two-digit number.

| ☐ | + | ☐ | = | ☐ | | ☐ | – | ☐ | = | ☐ |

| ☐ | + | ☐ | = | ☐ | | ☐ | – | ☐ | = | ☐ |

My Homework

Homework Helper Need help? connectED.mcgraw-hill.com

Use related facts to complete a fact family.

$6 + 4 = 10$
$4 + 6 = 10$
$10 - 4 = 6$
$10 - 6 = 4$

Helpful Hint
Each fact in a fact family uses the same three numbers.

Practice

Complete each fact family.

1.

___ + ___ = ___

___ + ___ = ___

___ − ___ = ___

___ − ___ = ___

2.

___ + ___ = ___

___ + ___ = ___

___ − ___ = ___

___ − ___ = ___

3. There are 8 apples on one tree.
There are 7 apples on another tree.
How many apples are there in all?

_____ apples

4. Sam has 7 toy trucks in one box. He
has one less than that number in
another box. How many toy trucks
does he have in all?

Beep! Beep!

_____ toy trucks

Vocabulary Check

Vocab
a b c

Circle the correct answer.

5. Which group shows a fact family?

$2 + 3 = 5$	$1 + 4 = 5$	$2 + 1 = 3$
$3 + 2 = 5$	$1 + 5 = 6$	$2 + 2 = 4$
$5 - 3 = 2$	$1 + 7 = 8$	$2 + 3 = 5$
$5 - 2 = 3$	$1 + 8 = 9$	$2 + 4 = 6$

Math at Home Have your child write the fact family for the numbers 6, 7, and 13.

86   Chapter 1 • Lesson 12

Two-Step Word Problems

Explore and Explain

Tools

Lesson 13

ESSENTIAL QUESTION
What strategies can I use
to add and subtract?

I'm a boat
chasing champ!

_____ dolphins

Teacher Directions: Use counters to solve the problem. Trace the counters to show
your work. There is a group of 12 dolphins following a boat. 3 more dolphins join the
group. Then 6 of the dolphins swim away. How many dolphins are still following the boat?

See and Show

Some word problems take two steps to solve.

Tim's class was having a pizza party. Tim brought 3 pizzas. Jenni brought 2 pizzas. The class ate 4 pizzas. How many pizzas do they have left?

_____ pizza

Solve each two-step word problem.

1. Marla bought 3 books on Monday and 5 books on Wednesday. She returned 4 books today. How many books does Marla have now?

_____ books

2. Carla collected 6 leaves but she lost 3 of them. She has 9 leaves at home. How many leaves does Carla have in all?

_____ leaves

Talk Math Explain how to solve a two-step word problem.

Name _____

On My Own

Solve each two-step word problem.

3. Candice has 6 pencils. She gives 2 away. Then she gets 2 pencils out of her bag. How many pencils does Candice have now?

_____ pencils

4. 8 birds land on a tree. 5 of the birds fly to another tree. Then 2 more fly away. How many birds are left in the tree?

_____ birds

5. Suzie saw 9 ladybugs on her porch. 4 flew away. Then 3 more ladybugs came. How many ladybugs are on the porch now?

_____ ladybugs

Problem Solving

6. 5 seashells were lying on the beach.
8 more shells washed up onto the beach.
Sammy picked up 4 shells. How many
shells were left on the beach?

_____ shells

7. David counted 6 chipmunks and 7
rabbits in his yard. 5 of the rabbits
hopped away. How many animals
can David see now?

Cotton tail

_____ animals

8. There were 5 toads in the pond. 3 more toads
jumped in. 2 toads jumped out of the pond.
How many toads are in the pond now?

_____ toads

HOT Problem Write a two-step word problem using the
number sentences $8 + 2 = 10$ and $10 - 4 = 6$.

_____

_____

_____

My Homework

Homework Helper **Need help?** connectED.mcgraw-hill.com

Tony bought 3 items for lunch. Delia bought 3 items for lunch. Sally had 4 items for lunch. How many items did they have in all?

$$+\begin{array}{c}3\\3\\\hline 6\end{array} \qquad +\begin{array}{c}6\\4\\\hline 10\end{array} \qquad \boxed{10}\ \text{items}$$

Practice

Solve each two-step word problem.

1. Camden has 3 marbles in his pocket. He finds 8 more. He gives 4 marbles to Jen. How many marbles does Camden have now?

   _____ marbles

2. Alexis collected 6 rocks. She already had 9 rocks. She lost 3 rocks. How many rocks does Alexis have now?

   _____ rocks

Solve each two-step word problem.

I wonder who lives in that house?

3. 9 geese landed in the yard. 8 of the geese flew away. Then 3 more geese landed in the yard. How many geese are in the yard now?

_____ geese

4. Sheila had 15 cards to mail to her friends. She forgot to mail 2 of them. 6 of them got lost in the mail. How many cards were delivered?

_____ cards

5. Gia got 4 books from the library yesterday. She got 5 books today. She returned 3 books. How many library books does Gia have now?

_____ books

Test Practice

6. There are 7 horses on the farm. There are also 9 pigs. 2 of the pigs were sent to another farm. How many animals are on the farm now?

18	14	7	5
○	○	○	○

Math at Home Tell a two-step number story to your child. Have him or her use pennies to show you the steps they use to solve it.

Name

Fluency Practice

No monkey business!

Add.

1. $1 + 8 =$ _____

2. $9 + 7 =$ _____

3. $3 + 8 =$ _____

4. $6 + 5 =$ _____

5. $7 + 5 =$ _____

6. $8 + 4 =$ _____

7. $8 + 5 =$ _____

8. $8 + 8 =$ _____

9. $7 + 4 =$ _____

10. $4 + 8 =$ _____

11. $3 + 5 =$ _____

12. $6 + 8 =$ _____

13. $7 + 9 =$ _____

14. $7 + 8 =$ _____

15. $7 + 7 =$ _____

16. $4 + 4 =$ _____

17. $4 + 3 =$ _____

18. $5 + 4 =$ _____

19. $8 + 8 =$ _____

20. $9 + 9 =$ _____

21. $7 + 6 =$ _____

22. $9 + 8 =$ _____

23. $8 + 9 =$ _____

24. $5 + 9 =$ _____

Fluency Practice

Subtract.

1. $8 - 1 =$ _____

2. $16 - 7 =$ _____

3. $8 - 3 =$ _____

4. $11 - 5 =$ _____

5. $7 - 5 =$ _____

6. $12 - 4 =$ _____

7. $8 - 5 =$ _____

8. $17 - 8 =$ _____

9. $7 - 4 =$ _____

10. $12 - 8 =$ _____

11. $5 - 3 =$ _____

12. $15 - 6 =$ _____

13. $16 - 9 =$ _____

14. $15 - 8 =$ _____

15. $14 - 7 =$ _____

16. $8 - 4 =$ _____

17. $7 - 3 =$ _____

18. $9 - 4 =$ _____

19. $16 - 8 =$ _____

20. $18 - 9 =$ _____

21. $13 - 6 =$ _____

22. $17 - 8 =$ _____

23. $17 - 9 =$ _____

24. $9 - 5 =$ _____

Name
_____

My Review

Vocabulary Check

add	addend	count back	count on
fact family	related facts	subtract	

Write the correct word in the blank.
Use words from the box.

1. When you take one part away from a whole, you
   _____.

2. Addition and subtraction facts that use the same three numbers can be called a _____
   or _____.

3. The two numbers you add together in an addition sentence are called _____.

4. You can _____ to subtract 2 from a number.

5. To _____, you join two parts together.

6. You can _____ to add 2 to a number.

Concept Check

Add.

7. 9
   + 3
   ___

8. 8
   + 5
   ___

9. 9
   + 6
   ___

10. 7
    + 7
    ___

Find each sum.

11. 6 + 5 + 4 = _____

12. 2 + 2 + 8 = _____

Use addition facts to subtract.

13. 7 + 8 = _____

    15 − 7 = _____

14. 6 + 7 = _____

    13 − 7 = _____

Complete the fact family.

15.

_____ + _____ = _____

_____ + _____ = _____

_____ − _____ = _____

_____ − _____ = _____

Name ....

Problem Solving

16. Dotty has 8 red flowers and 2 pink flowers. She gives 5 flowers to her mother. How many flowers does Dotty have left?

_____ flowers

17. Oscar finds 6 leaves in the morning. He finds 2 more leaves after lunch. He already has 8 leaves. How many leaves does he have in all?

_____ leaves

18. Shani walks the dogs in her neighborhood. One neighbor has 4 dogs. Another neighbor has 3 dogs. How many dogs does she walk?

_____ dogs

Test Practice

19. Rod throws 5 baseballs. Ty throws the same number of baseballs as Rod. How many baseballs did they throw in all?

0	5	10	15
○	○	○	○

Reflect

Show different strategies to add and subtract.

Make a 10.

$5 + 9 =$ _____

$10 +$ _____ $=$ _____

Use doubles to add
and subtract.

$7 + 7 =$ _____

$14 - 7 =$ _____

ESSENTIAL
QUESTION
**What strategies
can I use to add
and subtract?**

Complete the
fact family.

2 + 5 = 7

_____ + _____ = _____

_____ - _____ = _____

_____ - _____ = _____

Find the missing addend.

$3 +$ _____ $= 9$

$9 - 3 =$ _____

I knew
you could!

See, you
can do it!

ESSENTIAL QUESTION
How can equal groups help me add?

We're Going to the Desert

Watch a video!

Watch

Chapter 2 Project

Math Posters

1. Work with your group to write a math concept below. Then define or draw a picture of the concept. You may use each of your group members' pages to show different concepts.

2. Glue all of your group's pages onto the poster board.

3. Present your Math Poster to the class.

Name _____

Am I Ready?

1. Write the missing numbers.

0 1 ☐ 3 ☐ 5 ☐ ☐ 8 ☐ 10

Write the missing numbers.

2. _____, 2, _____, _____, 5, 6, _____

3. 11, _____, _____, 14, _____, 16, _____

4. 10, 20, _____, 40, _____, 60, _____

Add.

5. 5 + 5 = _____   6. 2 + 2 + 2 = _____

7. Alisa has 3 red cups, 3 blue cups, and
    3 pink cups. How many cups does she

    have in all? _____

Shade the boxes to show the problems you answered correctly.

 How Did I Do?

| 1 | 2 | 3 | 4 | 5 | 6 | 7 |

My Math Words

Review Vocabulary

addend   sum

Write a review word in each box to describe the parts of each addition number sentence.

7   +   9   =   16

[ ]   [ ]   [ ]

13   =   6   +   7

[ ]   [ ]   [ ]

5   +   8   =   13

[ ]   [ ]   [ ]

My Vocabulary Cards

Lesson 2–5

array

2 groups of 4

Lesson 2–2

equal groups

3 equal groups

Lesson 2–6

even

1, 3, 5,
17, 45, 89

2, 4, 6,
24, 52, 60

Lesson 2–6

odd

1, 3, 5,
17, 45, 89

2, 4, 6,
24, 52, 60

Lesson 2–4

repeated addition

$$\begin{array}{r} 3 \\ 3 \\ 3 \\ +\underline{3} \\ 9 \end{array}$$

Lesson 2-1

skip count

5, 10, 15, 20, 25

+5   +5   +5   +5

Teacher Directions:
Ideas for Use
• Group 2 or 3 common words. Add a word that is unrelated to the group. Ask another student to name the unrelated word.

• Arrange the cards in alphabetical order.

Each group has the same number of objects.

Objects displayed in rows and columns.

An odd number of objects has 1 left over when counted by 2s. Numbers that end with 1, 3, 5, 7, or 9.

An even number of objects can be counted by 2s. Numbers that end with 0, 2, 4, 6, 8.

Count objects in equal groups of two or more.

To use the same addend over and over.

My Foldable

 FOLDABLES® Follow the steps on the back to make your Foldable.

_____ rows of _____

$3 + 3 =$ _____

_____ rows of _____

_____ = _____

Repeated Addition with

Arrays

5 rows of 4

_____ = _____

_____ rows of _____

_____ = _____

Skip Count On a Hundred Chart

Lesson 1

ESSENTIAL QUESTION
How can equal groups help me add?

Explore and Explain

1		3		5		7		9	
11		13		15		17		19	
21		23		25		27		29	
31		33		35		37		39	
41		43		45		47		49	
51		53		55		57		59	
61		63		65		67		69	
71		73		75		77		79	
81		83		85		87		89	
91		93		95		97		99	

Let's start coloring!

Teacher Directions: Count by 2s. Write the missing numbers. Count by 5s. Color the numbers yellow. Count by 10s. Circle those numbers red.

See and Show

Processes & Practices

Use patterns to count. You can **skip count** by 2s, 5s, and 10s.

Helpful Hint
Skip counting on a hundred chart creates patterns.

1	2	3	4	5	6	7	8	9	10
11	12	13	14	15	16	17	18	19	20
21	22	23	24	25	26	27	28	29	30
31	32	33	34	35	36	37	38	39	40
41	42	43	44	45	46	47	48	49	50
51	52	53	54	55	56	57	58	59	60

Skip count by 2s: 2, 4, __6__, 8, __10__, 12

Skip count by 5s: 5, 10, __15__, 20, 25, __30__

Skip count by 10s: 10, 20, __30__, 40, __50__, __60__

Skip count. Write the missing numbers.

1. Skip count by 2s: 6, _____, 10, _____, 14, _____

2. Skip count by 5s: 15, 20, _____, 30, _____, 40

3. Skip count by 10s: _____, 50, 40, _____, 20, _____

Talk Math   When would you use skip counting?

108   Chapter 2 • Lesson 1

Copyright © The McGraw-Hill Companies, Inc.

Name _____

On My Own

Use the hundred chart to skip count.

1	2	3	4	5	6	7	8	9	10
11	12	13	14	15	16	17	18	19	20
21	22	23	24	25	26	27	28	29	30
31	32	33	34	35	36	37	38	39	40
41	42	43	44	45	46	47	48	49	50
51	52	53	54	55	56	57	58	59	60
61	62	63	64	65	66	67	68	69	70
71	72	73	74	75	76	77	78	79	80
81	82	83	84	85	86	87	88	89	90
91	92	93	94	95	96	97	98	99	100

Skip counting is hot!

4. Start on 5. Count by 5s. Color the numbers blue.

5. Start on 10. Count by 10s. Circle the numbers red.

Write the missing number. Describe the pattern.

6. 34, 36, 38, _____

Skip count by _____.

7. 10, _____, 30, 40

Skip count by _____.

8. 24, 22, _____, 18

Skip count by _____.

9. 40, 35, _____, 25

Skip count by _____.

10. If there are 2 vases with 10 flowers in each vase, how many flowers are there in all?

_____ flowers

11. There are 3 fish bowls. How many fish are there in all if there are 5 fish in each bowl?

_____ goldfish

12. How many frogs are there in all if there are 3 lily pads and each one has 2 frogs?

_____ frogs

 Write an addition story for the cubes.

_____

_____

_____

My Homework

Homework Helper Need help? connectED.mcgraw-hill.com

You can skip count on a hundred chart.

Skip count by 2s.
58, 60, 62, 64, 66, 68

Skip count by 5s.
55, 60, 65, 70, 75, 80

Skip count by 10s.
60, 70, 80, 90, 100

51	52	53	54	55	56	57	58	59	60
61	62	63	64	65	66	67	68	69	70
71	72	73	74	75	76	77	78	79	80
81	82	83	84	85	86	87	88	89	90
91	92	93	94	95	96	97	98	99	100

Practice

Skip count. Write the missing numbers.

1. 62, _____, 66, _____, _____, 72

2. 60, _____, 70, _____, 80, _____

3. 95, _____, _____, 80, 75, _____

4. 88, _____, 84, _____, _____, 78

Write the missing number. Describe the pattern.

5. 50, _____, 70, 80

   Skip count by _____.

6. 30, 35, _____, 45

   Skip count by _____.

7. 75, _____, 85, 90

   Skip count by _____.

8. 90, _____, _____, 60

   Skip count by _____.

9. **Draw a picture to solve.**

   There are 20 wheels on a group of bicycles. Each bike has 2 wheels. How many bikes are there?

   _____ bicycles

Vocabulary Check

10. Which is not an example of a way to **skip count**? Circle it.

    10, 12, 16, 20, 24   15, 20, 25, 30, 35

    28, 26, 24, 22, 20   40, 50, 60, 70, 80

Adventure time!

Math at Home Practice skip counting by 2s, 5s and 10s. Use cereal or beans.

Skip Count by 2s, 5s, and 10s

Lesson 2
ESSENTIAL QUESTION
How can equal groups help me add?

Explore and Explain

This desert is not big enough for all of us.

_____ , _____ , _____ snakes

 Teacher Directions: Use counters to model. There are 3 groups of snakes in the desert. There are 5 snakes in each group. Skip count. Write the numbers. How many snakes in all?

See and Show

Equal groups have the same number of objects. Put two counters in each group. How many equal groups are there?

_____ 4 _____ equal groups

Skip count to find the total.

> **Helpful Hint**
> When there are equal groups, you can skip count to find the total.

_____2 , _____4 , _____6 , _____8 counters in all

Use ⬤ to model equal groups.
Skip count to find the total.

1. 3 groups of 2 _____ in all

2. 4 groups of 5 _____ in all

Describe the groups. Skip count to find the total.

3.

_____ groups of _____   _____ in all

Talk Math   Create a skip-counting story for Exercise 3.

Name _____

On My Own

Use to model equal groups.
Skip count to find the total.

4. 2 groups of 10   _____ in all

5. 6 groups of 2   _____ in all

Describe the groups. Skip count to find the total.

6.

_____ groups of _____   _____ in all

7.

_____ groups of _____   _____ in all

8.

_____ groups of _____   _____ in all

Problem Solving

Solve. Draw a picture to help, if needed.

9. Lori has 6 bunches of grapes.
   Each bunch has 10 grapes.
   How many grapes does Lori
   have in all?

   _____grapes

10. Paul organized his shells in groups of 5.
    He has 3 groups of 5 shells. How many shells
    does he have in all?

    _____ shells

11. Myla put 10 cookies each in 4 bags for a bake
    sale. Kate put 10 cookies in each of 3 bags.
    How many cookies do they have in all?

    _____ cookies

Write Math How can you find the total number of
tennis balls in 4 groups of 10 tennis balls?
Use your vocabulary word.

_____

_____

_____

Name _____

My Homework

Homework Helper eHelp   Need help? connectED.mcgraw-hill.com

Make equal groups and skip count to find the total.

3 equal groups

2   4   6   owls in all

Practice

Skip count to find the total.

1.

_____ groups of _____   _____ spiders in all

2.

_____ groups of _____   _____ turtles in all

Solve. Draw a picture to help, if needed.

3. Evan has 6 groups of 5 toy cars. How many toy cars does he have in all?

_____ toy cars

4. Mrs. Hanes is planning a field trip. There are 6 groups of 10 students going. How many students are going on the field trip?

_____ students

5. Austen has 6 sets of 10 baseball cards. Tim has 3 sets of 10 baseball cards. How many baseball cards do they have in all?

_____ baseball cards

Vocabulary Check

6. Circle **equal groups** of five buttons.

Math at Home Use buttons, macaroni, or pennies. Have your child make equal groups of 2s, 5s, and 10s. Have him or her skip count to find the totals.

Problem Solving

STRATEGY: **Find a Pattern**

Lesson 3

ESSENTIAL QUESTION
How can equal groups help me add?

Kelly puts 5 party favors in each bag. She has 6 bags. How many favors does she need in all?

Watch ▶ Tools

Party time!

1 Understand
Underline what you know.
Circle what you need to find.

Giddy up!

2 Plan
How will I solve the problem?

3 Solve
Find a pattern.

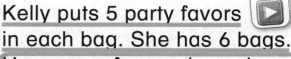

5 ,	10 ,	15 ,	20 ,	25 ,	30
bag 1	bag 2	bag 3	bag 4	bag 5	bag 6

She needs __30__ favors in all.

4 Check
Is my answer reasonable? Explain.

Practice the Strategy

Kate and Pedro count the tickets they sold to attend a museum. They sold 10 tickets each day for 5 days. How many tickets were sold?

Let's go to the Wild West Museum!

1 Understand Underline what you know.
Circle what you need to find.

2 Plan How will I solve the problem?

3 Solve I will . . .

_____, _____, _____, _____, _____

By day 5, _____ tickets were sold in all.

4 Check Is my answer reasonable? Explain.

Apply the Strategy

1. Xavier is looking at a map of the desert. He knows that each finger width is about 10 miles. How many miles will he count if he uses 7 finger widths?

_____ miles

2. Sam thinks of the number pattern 15, 20, 25, 30. He continues this pattern. What will be the next four numbers?

_____, _____, _____, _____

3. Amar sees this number pattern. What is the missing number?

10, _____, 20, 25, 30

Review the Strategies

Choose a strategy
- Write a number sentence.
- Act it out.
- Find a pattern.

4. Justin is stacking books. There are 5 books in the first stack. 10 in the second stack. The third stack has 15 books. The pattern continues. How many books will be in the next stack?

_____ books

5. On the first day of the food drive, Ms. Buckle's class collects 8 cans. Mr. Cline's class collects 6 cans, and Mrs. Brown's class collects 5 cans. How many cans do they collect in all?

_____ cans

6. Josh recorded how many inches of snow fell in each month. Continue the pattern. How many inches fell in January?

Month	Inches
November	14
December	16
January	
February	20

_____ inches

My Homework

Deb paints 2 flowers in the first row, 4 flowers in the second row, and 6 flowers in the third row. If she continues the pattern, how many flowers will be in the fifth row?

1 Understand
Underline what you know.
Circle what you need to find.

2 Plan
How will I solve the problem?

It's a masterpiece!

3 Solve
I will find a pattern.

2,   4,   6,   8,   10
Row 1, Row 2, Row 3, Row 4, Row 5

There are 10 flowers in the fifth row.

4 Check
Is my answer reasonable?

Practice

Underline what you know. Circle what you need to find.
Find a pattern to solve.

1. One elephant has four legs.
   Two elephants have eight legs.
   How many legs do five elephants have? _____

2. Corey is thinking of a number.
   It is the next number in the pattern
   50, 45, 40, 35. What is the number? _____

3. Brad has groups of 10 pennies.
   Brad counts 40, but he has 6 more
   stacks to count. How many pennies
   does he have?

   _____ pennies

Test Practice

4. Which of the following is an example of skip counting?

   8, 10, 20, 30, 40   40, 35, 25, 15, 20
           ○   ○

   20, 25, 30, 35, 40   45, 55, 60, 70, 80
           ○   ○

Math at Home Ask your child to write a number pattern that uses 5 or 6 numbers.
Have him or her explain the pattern to you.

Name

Check My Progress

Vocabulary Check

Complete each sentence.

skip count   **equal groups**

1. When you have sets of the same number of objects you have _____.

2. You can _____ to help you count equal groups.

Concept Check

Skip count. Write the missing numbers.

3. 6, _____, _____, 12, 14, _____, 18, _____

4. 15, 20, _____, 30, _____, 40, _____, 50

5. 20, _____, _____, 50, _____, _____, 80, 90

6. _____, 55, _____, 65, _____, _____, 80

7. 15, 20, _____, 30

   Skip count by _____.

8. 20, _____, 16, 14

   Skip count by _____.

Describe the groups. Skip count to find the total.

9.

_____ groups of _____   _____ ants in all

10.

_____ groups of _____   _____ lizards in all

---

11. Jason has 3 groups of 5 baseball cards.
Cameron has 2 groups of 10 baseball cards.
Which boy has more cards?

_____

Test Practice

12. Jose has 3 boxes of rocks in his collection. Each box holds 10 rocks. He fills one more box. How many rocks does he have in all?

10	13	30	40
○	○	○	○

Repeated Addition

Explore and Explain

 Watch Tools

_____ + _____ + _____ = _____

 Teacher Directions: Place a connecting cube on each blade of the wind turbines. Write how many cubes are on each turbine. Count all the connecting cubes. How many blades in all? Write the total for the number sentence.

Online Content at ⌐ **connectED.mcgraw-hill.com**

See and Show

When groups are equal, you can use **repeated addition** to find the total.

> **Helpful Hint**
> You can skip count to find sums like this.

2 + 2 + 2 + 2 = __8__

3 + 3 + 3 = __9__

Use connecting cubes to model equal groups. Add.

1.

___ + ___ = ___

2.

___ + ___ + ___ = ___

3.

___ + ___ = ___

Talk Math Create a story for 2 + 2 + 2 + 2 + 2.

Name _____

On My Own
Add.

4.

___ + ___ + ___ = ___

5.

___ + ___ + ___ + ___ = ___

6.

___ + ___ + ___ = ___

7. Draw your own example. Then add.

4 + 4 + 4 + 4 = ___

Problem Solving

Processes & Practices

Use repeated addition to solve.

8. Mike has five pairs of socks. Each pair has two socks. How many socks does Mike have?

_____ socks

9. Brad makes four groups of animal cards. Each group has three cards. How many cards does he make?

_____ cards

10. There are four balloons in each bunch. Marcy has four bunches. How many balloons are there in all?

Oh, no! a cactus!

_____ balloons

HOT Problem Jaya writes a repeated number sentence. It has three numbers. The sum is 15. What is the number sentence? Explain.

_____

_____

_____

Copyright © The McGraw-Hill Companies, Inc. Ingram Publishing/Superstock

Name

....

My Homework

Homework Helper Need help? connectED.mcgraw-hill.com

When groups are equal, use repeated addition
to find the total.

2   +   2   +   2   = 6 cactus

Practice

Add.

1.

_____ + _____ + _____ = _____

2.

_____ + _____ + _____ = _____

Chapter 2 • Lesson 4   131

Add.

3.

_____ + _____ = _____lady bugs

Use repeated addition to solve.

4. Marco has 4 fish tanks. Each tank has 10 fish. How many fish does Marco have in all?

_____ fish

5. Mandy has 4 boxes of raisins. Each box has 5 raisins. How many raisins does Mandy have?

_____ raisins

Look! I made "Ants on a Log!"

Vocabulary Check

6. Circle the **repeated addition** sentence.

5, 10, 15, 20

9 + 3 + 9

3 + 3 + 3

 Math at Home Have your child put beans or pennies in 5 equal groups of 6 to show repeated addition. Ask them to write a number sentence and solve.

Repeated Addition with Arrays

Explore and Explain

 Tools

We love collecting rocks!

That's easy for her to say. Her rock is the smallest!

Jamar

Meko

Kyle

Laurel

____ + ____ + ____ + ____ = ____ rocks

Teacher Directions: Jamar and 3 of his friends each collect 2 rocks. How many rocks do they have in all? Use connecting cubes to show equal groups of rocks. Color in the grid to show your work. Write a number sentence to show the total.

See and Show

In an **array**, objects are shown in rows and columns.

3 rows of 4

3 rows of 4

$$\underline{4} + \underline{4} + \underline{4} = \underline{12}$$

Describe each array using a number sentence.

1.

___ + ___ + ___ = ___

____ rows of ____ turtles

2.

___ + ___ = ___

____ rows of ____ cacti

3. Shade the grid to show 4 rows of 2.
   Write a number sentence to describe it.

   _____

Talk Math   How can arrays help you add?

Name _____

We're 1 group of 3!

On My Own

Describe the array using a number sentence.

4.

___ + ___ + ___ = ___   ____ rows of ____ lizards

Shade each grid to show the array.
Write a number sentence to describe it.

5. Show 4 rows of 3.

_____

6. Show 4 rows of 1.

_____

7. Show 2 rows of 5.

_____

8. Show 3 rows of 4.

_____

Copyright © The McGraw-Hill Companies, Inc. Blend Images/PunchStock

Use an array to solve.

9. Kathy puts 5 chairs in one row. She puts the same number of chairs in three more rows. How many chairs are there in all?

I call dibs on this one.

_____ chairs

10. Three hills each have 4 wind turbines. How many wind turbines are there in all?

_____ wind turbines

Write Math   Describe how this array shows the number sentence 2 + 2 + 2 = 6.

_____

_____

_____

Name

My Homework

Homework Helper Need help? connectED.mcgraw-hill.com

The **array** shows 5 rows of 3.

$$3 + 3 + 3 + 3 + 3 = 15$$

Practice

Describe each array using a number sentence.

1.

___ + ___ + ___ = ___

___ rows of ___ mice

2.

___ + ___ = ___

___ rows of ___ owls

3. Show 3 rows of 5.

_____

4. Show 3 rows of 2.

_____

Describe each array using a number sentence.

Hoorah for arrays!

5. Show 3 rows of 5.

_____

6. Show 4 rows of 2.

_____

7. Show 2 rows of 4.

_____

8. There are 4 legs on 1 camel. How many legs are on 5 camels?

_____ legs in all

Vocabulary Check

9. Draw an **array** of 2 rows of 3.

Math at Home Have your child think of things around them that are in an array. Ex: eggs in a carton, days on the calendar, etc.

Name ....

Even and Odd Numbers

Remember, leftovers mean it's odd!

Explore and Explain ▶Watch ◉Tools

Try to break the cubes into two equal groups.

If the groups are equal with none left over, your number is even.

If you have one left over cube, your number is odd.

 Teacher Directions: Roll a blue number cube. Make a cube train to show the number. Try to break the train into two equal groups. Place any extras on the red cube. Is the number even or odd? Repeat. Draw the groups.

Online Content at ⬐**connectED.mcgraw-hill.com**   Chapter 2 • Lesson 6   139

See and Show

An **even** number of objects can be counted by 2s.

4   2   4

> **Helpful Hint**
> Start with 4 cubes.
> Make pairs. Count. 2, 4

An **odd** number of objects has 1 left over when counted by 2s.

5   2   4   5

> Start with 5 cubes.
> Make pairs. 1 is left over. Count. 2, 4, 5

**Use connecting cubes to show the number.
Circle *even* or *odd*.**

1.
4   even
       odd

2.
5   even
       odd

3.
8   even
       odd

4.
9   even
       odd

Write the number of cubes. Circle even or odd.

5. _____ even   odd

6. _____ even   odd

Talk Math   When might you use even and odd numbers?

Name _____

On My Own

Use connecting cubes to show the number. Circle *even* or *odd*.

7.   11   even
           odd

8.   12   even
           odd

9.   17   even
           odd

10. 14   even
           odd

11. 8   even
           odd

12. 19   even
           odd

Write the number of cubes. Circle *even* or *odd*.

13.

_____ even   odd

14.

_____ even   odd

15.

_____ even   odd

16.

_____ even   odd

17.

_____ even   odd

18. Ben has 16 pennies. Does he have an even or odd number of pennies?

_____

19. Mr. Rice thinks of a number between 14 and 17. The number is even. What is the number?

_____

20. Miss Lee has 18 plants. She wants to split them between two friends. How many plants will she give each friend?

I think it's time for a haircut

_____ plants

Write Math
Explain how you know 15 is an odd number.

_____

_____

_____

Name _____

My Homework

Homework Helper Need help? connectED.mcgraw-hill.com

An even number of objects can be counted by 2s.

Helpful Hint
Make pairs.
Circle them.

4   2   4

An odd number of objects has 1 left over when counted by 2s.

5   2   4   5

Practice

Use buttons or pennies to show each number.
Circle _even_ or _odd_.

1. 8	even / odd
2. 16	even / odd
3. 11	even / odd
4. 19	even / odd

Write the number of buttons. Circle *even* or *odd*.

Helpful Hint
Circle pairs. If none are left over, the number is even.

5.

_____   even   odd

6.

_____   even   odd

Red Rover! Red Rover! Let Billy come over!

7.

_____   even   odd

8. Katy has 13 oranges, 12 apples, and 6 bananas. She also has two boxes. Which fruit cannot be put evenly into the two boxes?

_____

Vocabulary Check

Draw lines to match.

9. **odd**   A number that can be counted by 2s.

10. **even**   A number that can be counted by 2s and has one left over.

Math at Home Give your child random numbers up to 20 and let them tell you if they are even or odd. Use buttons or coins if needed.

Name ....

Sums of Equal Numbers

Lesson 7

ESSENTIAL QUESTION
How can equal groups help me add?

Explore and Explain

Tools

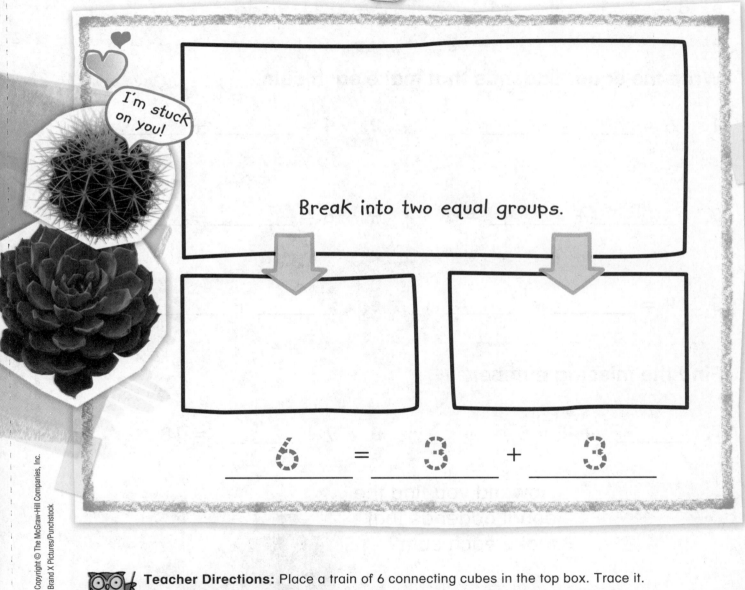

I'm stuck on you!

Break into two equal groups.

6 = 3 + 3

Teacher Directions: Place a train of 6 connecting cubes in the top box. Trace it. Break it into two equal groups. Put one group in each of the two bottom boxes. Trace them. Trace the number sentence.

See and Show

Any even number can be written as the sum of two equal addends.

$8 = \underline{4} + \underline{4}$

$2 = \underline{1} + \underline{1}$

Write the equal addends that make each sum.

1. $6 = \underline{\qquad} + \underline{\qquad}$

2. $4 = \underline{\qquad} + \underline{\qquad}$

3. $\underline{\qquad} + \underline{\qquad} = 2$

4. $8 = \underline{\qquad} + \underline{\qquad}$

5. $14 = \underline{\qquad} + \underline{\qquad}$

6. $\underline{\qquad} + \underline{\qquad} = 18$

Find the missing number.

7. $\underline{\qquad} + 4 = 8$

8. $9 + \underline{\qquad} = 18$

Talk Math How did you find the equal addends that make each sum?

Name _____

On My Own

Write the equal addends that make each sum.

9. 2 = _____ + _____

10. 12 = _____ + _____

11. 10 = _____ + _____

12. 14 = _____ + _____

13. 18 = _____ + _____

14. 8 = _____ + _____

15. _____ + _____ = 6

16. _____ + _____ = 4

17. _____ + _____ = 20

18. _____ + _____ = 16

Find the missing addends.

19. 10 + _____ = 20

20. _____ + 6 = 12

21. 4 + _____ = 8

22. _____ + 3 = 6

23. _____ + 7 = 14

I may live in the desert, but I'm a cool cat!

Problem Solving

24. Henry buys 6 apples. His mother buys the same amount. Write the number sentence. How many apples do they have?

_____ + _____ = _____

25. Luke has 10 puppies. An equal number of them are black and brown puppies. How many are brown?

_____ brown puppies

26. Janelle brought 7 magazines. The class now has 14 magazines. How many magazines did they already have?

_____ magazines

Write Math   Tamika and Jen have the same number of beads. They have 18 beads in all. How many beads does each girl have? Explain.

_____

_____

_____

Name _____

My Homework

Homework Helper   Need help? ⬚ connectED.mcgraw-hill.com

Any even number can be written as the sum
of two equal addends.

6 =   3   +   3   4 = 2 + 2

Practice

Write the equal addends that make each sum.

1. 10 = _____ + _____

2. 2 = _____ + _____

3. 18 = _____ + _____

4. 8 = _____ + _____

Find the missing number.

5. 7 + _____ = 14

6. _____ + 3 = 6

7. 8 + _____ = 16

8. _____ + 6 = 12

Find the missing addends.

9. $5 + \underline{\hspace{1cm}} = 10$

10. $\underline{\hspace{1cm}} + 2 = 4$

11. Maria and Steve each have 8 toys for the toy drive. How many toys do they have in all?

I couldn't eat another thing. I'm stuffed!

$\underline{\hspace{1cm}}$ toys

12. Mrs. Miller's class has 18 students. There are an equal number of boys and girls in the class. Write a number sentence to show how many boys and how many girls are in the class.

$\underline{\hspace{1cm}} = \underline{\hspace{1cm}} + \underline{\hspace{1cm}}$

Test Practice

13. Kylie buys 7 bananas. Her mom also buys 7 bananas. How many bananas do they buy in all?

0	7	10	14
◯	◯	◯	◯

Math at Home Work with your child to create a number line from 1 to 20. Have your child circle 18, 14, and 10. Ask your child to tell you the equal addends for each of these numbers.

Fluency Practice

Add.

1. $2 + 8 =$ _____

2. $8 + 4 =$ _____

3. $3 + 6 =$ _____

4. $5 + 5 =$ _____

5. $6 + 5 =$ _____

6. $6 + 4 =$ _____

7. $6 + 6 =$ _____

8. $7 + 7 =$ _____

9. $7 + 6 =$ _____

10. $4 + 2 =$ _____

11. $4 + 7 =$ _____

12. $5 + 8 =$ _____

13. $8 + 9 =$ _____

14. $5 + 7 =$ _____

15. $8 + 8 =$ _____

16. $2 + 4 =$ _____

17. $2 + 3 =$ _____

18. $3 + 4 =$ _____

19. $5 + 8 =$ _____

20. $8 + 9 =$ _____

21. $3 + 5 =$ _____

22. $9 + 9 =$ _____

23. $0 + 9 =$ _____

24. $4 + 8 =$ _____

Fluency Practice

Subtract.

1. 10 − 1 = _____

2. 9 − 7 = _____

3. 9 − 3 = _____

4. 11 − 4 = _____

5. 6 − 5 = _____

6. 13 − 6 = _____

7. 7 − 4 = _____

8. 10 − 8 = _____

9. 11 − 4 = _____

10. 14 − 7 = _____

11. 5 − 1 = _____

12. 15 − 6 = _____

13. 17 − 9 = _____

14. 11 − 8 = _____

15. 16 − 7 = _____

16. 12 − 5 = _____

17. 12 − 3 = _____

18. 9 − 6 = _____

19. 18 − 9 = _____

20. 15 − 7 = _____

21. 13 − 7 = _____

22. 17 − 9 = _____

23. 17 − 8 = _____

24. 9 − 8 = _____

Name _____

My Review

Vocabulary Check

array	equal groups	even
odd	repeated addition	skip count

Complete each sentence.

1. _____ have the same number of objects.

2. You can _____ by counting objects in equal groups of two or more.

3. An _____ number of objects has 1 left over when counted by 2s.

4. An _____ is a group of objects arranged in rows and columns of the same length.

Concept Check

Write the missing numbers.

5. 65, _____, 75, 80, _____, 90, _____, 100

6. 30, _____, 50, 60, _____, 80, _____

Fill in the blanks to describe each array.

7.

_____ rows of _____

8.

_____ rows of _____

Circle _even_ or _odd_.

9.   8   even
          odd

10. 13   even
          odd

11. 10   even
          odd

Write the equal addends that make each sum.

12. 8 = _____ + _____

13. 16 = _____ + _____

14. 12 = _____ + _____

15. 18 = _____ + _____

Find the missing addends.

16. 7 + _____ = 14

17. _____ + 9 = 18

18. _____ + 6 = 12

19. 3 + _____ = 6

20. _____ + 5 = 10

21. 7 + _____ = 14

Name _____

22. Liam has 5 dogs. Each dog has 4 bones. How many bones do the dogs have in all?

_____ bones

23. Lauren has 3 pages of stamps. Each page has 5 stamps on it. How many stamps does Lauren have in all?

_____ stamps

24. Jonah has an even number of golf balls. He has more than 13 and less than 16 golf balls. How many golf balls does he have?

_____ golf balls

Test Practice

25. Katherine has skip counted 16 rubber bracelets by 2s. If she continues skip counting by 2s, how many does she have after counting 5 more sets?

20	21	24	26
○	○	○	○

Reflect

Show how equal groups help you.

Write a repeated
addition sentence that
describes the cubes.

_____

Shade the array to show
4 rows of 5.

**ESSENTIAL
QUESTION**

**How can equal
groups help me add?**

Circle pairs.

Skip count by 5s. What are
the missing numbers?

40, _____, _____, 55

Is the number of
counters even
or odd?

_____

You can go far!

ESSENTIAL QUESTION

How can I add two-digit numbers?

I Like Teamwork!

Watch a video!

Watch

157

Chapter 3 Project

A Guide to Adding Two-Digit Numbers

1. After each lesson, create a page that shows how to solve addition problems.
2. Your group may want to include definitions, vocabulary, create examples, and write steps to solve problems.
3. Create a cover for your guide below.

Name _____

Am I Ready?

Write each number two ways.

1. 18 ones = _____

   _____ ten _____ ones

2. 26 ones = _____

   _____ tens _____ ones

Add.

3.   2
   + 7
   ____

4.   3
   + 4
   ____

5.   2
   + 2
   ____

6.   9
   + 6
   ____

7. 3 + 3 + 1 = _____

8. 4 + 2 + 3 = _____

9. Jack has 5 red marbles. Trent has 3 blue
   marbles. Lien has 2 green marbles. How
   many marbles do they have altogether?

   _____ marbles

**Shade the boxes to show the
problems you answered correctly.**

1	2	3	4	5	6	7	8	9

My Math Words

Review Vocabulary

addends	sum	add

Use the review words to complete the graphic organizer.

$$5 + 6 = 11$$

The blue number is called a _____.

The red numbers are called _____.

To join together to find the total is to _____.

Write a word problem about the addition sentence above.

_____

_____

_____

My Vocabulary Cards

Vocab
abc

Processes
&Practices

Lesson 3–2

regroup

18 + 5 = 23

Directions:
Ideas for Use

- Have students write a tally mark on the card each time they read the word in this chapter.

- Ask students to use the blank cards to write their own vocabulary cards.

Take apart a number
to write it in a new way.

✂ ---

$15 + 28 =$

$$\begin{array}{r} 22 \\ +\ 36 \\ \hline \end{array}$$

$54 + 37 =$

$$\begin{array}{r} 45 \\ +\ 19 \\ \hline \end{array}$$

$82 + 16 =$

$$\begin{array}{r} 31 \\ +\ 47 \\ \hline \end{array}$$

tens

ones

Name _____

Take Apart Tens to Add

Explore and Explain

We'll win if we pull together!

＿＿＿ + ＿＿＿ = ＿＿＿

 Teacher Directions: Use base-ten blocks. Show 48 in the red box. Show 16 in the yellow box. Move some base-ten blocks from the yellow box to the red box to make 50. Write the number shown in each box. Write the number sentence.

See and Show

Numbers that end in zero are easier to add.
Take apart addends to make numbers that end in zero.

Find 28 + 36.

One Way:

Take apart 36 as 2 + 34.
Then make 28 + 2 into 30.

28 + 36

28 + 2 + 34

30 + 34 = __6⁴__

Another Way:

Take apart 28 as 24 + 4.
Then make 4 + 36 into 40.

28 + 36

24 + 4 + 36

24 + 40 = __6⁴__

So, 28 + 36 = __6⁴__.

Take apart an addend to solve.

1. 18 + 35

18 + __2__ + __33__

__20__ + __33__ = __53__

So, 18 + 35 = __53__.

2. 44 + 26

44 + ____ + ____

____ + ____ = ____

So, 44 + 26 = ____.

Talk Math How do you decide which addend to take apart when adding?

Name _____

On My Own

Take apart an addend to solve.

3.   36 + 45

36 + _____ + _____

_____ + _____ = _____

So, 36 + 45 = _____.

4.   25 + 58

_____ + _____ + 58

_____ + _____ = _____

So, 25 + 58 = _____.

5.   67 + 26

67 + _____ + _____

_____ + _____ = _____

So, 67 + 26 = _____.

6.   38 + 14

38 + _____ + _____

_____ + _____ = _____

So, 38 + 14 = _____.

Problem Solving

7. 28 people came to the bake sale in the morning. 34 people came to the bake sale in the afternoon. How many people came in all?

My granola bars are a hit!

_____ people

8. Bill collected 16 toys for the toy drive. Lisa collected 37 toys. How many toys did Bill and Lisa collect in all?

She's right. I'm yummy!

_____ toys

Write Math Explain how you would find 65 + 18 by taking apart an addend.

_____

_____

_____

Name _____

My Homework

Homework Helper Need help? ➚ connectED.mcgraw-hill.com

You can take apart numbers to add.

29 + 52

29 + 1 + 51

30 + 51 = 81

29 + 52

21 + 8 + 52

21 + 60 = 81

Helpful Hint
Numbers ending in zero are easier to add. Try to take apart an addend to make tens. Then add.

So, 29 + 52 = 81.

Practice

Take apart an addend to solve.

1.   35 + 27

35 + _____ + _____

_____ + _____ = _____

So, 35 + 27 = _____.

2.   48 + 23

48 + _____ + _____

_____ + _____ = _____

So, 48 + 23 = _____.

Take apart an addend to solve.

3. 36 + 55

36 + _____ + _____

_____ + _____ = _____

So, 36 + 55 = _____.

4. 17 + 68

17 + _____ + _____

_____ + _____ = _____

So, 17 + 68 = _____.

5. There were 38 winter hats and 36 scarves donated to a clothing drive. How many hats and scarves were donated in all?

Stick with me scarf. We make a good team!

_____ hats and scarves

Test Practice

6. How could you take apart 18 + 36 to find the sum?

18 + 36 = _____

9 + 9 + 36 ○

18 + 2 + 34 ○

1 + 8 + 36 ○

18 + 3 + 6 ○

Math at Home Ask your child to solve 49 + 13 by taking apart an addend.

Regroup Ones as Tens

Explore and Explain

Watch ▶ Tools

We make a great pair!

Fresh Lemonade 50¢

tens	ones

We make a great pair too!

We're not pears. We're lemons!

19 + 3 = _____

Also, _____ tens

and _____ ones.

 Teacher Directions: Show the number 19 in the chart using base-ten blocks. In the ones column, below the first number, show 3 more ones cubes. Count all of the ones. Make a group of 10. Trade it for a tens rod and place the rod in the tens column. Write the sum. Write how many tens and ones.

Copyright © The McGraw-Hill Companies, Inc. (l)SW Productions/Photodisc/Getty Images, (r)Ocean/CORBIS

Find 27 + 5.

Step 1
Use and ▪ to show 27 and 5.

tens	ones

Step 2
If there are ten or more ones, **regroup** 10 ones as 1 ten.

tens	ones

Step 3
Write the number of tens and ones.

tens	ones

3 tens 2 ones

27 + 5 = __32__

Use Work Mat 6 and base-ten blocks to add.

	Add the ones. Add the tens.	Do you regroup?	Write the sum.
1. 15 + 8	____ ten ____ ones	yes no	
2. 23 + 6	____ tens ____ ones	yes no	

 Talk Math   How do you know if you need to regroup?

Name

On My Own

Use Work Mat 6 and base-ten blocks to add.

Remember 10 ones equal 1 ten.

I will!

	Add the ones. Add the tens.	Do you regroup?	Write the sum.
3. 76 + 4	____ tens ____ ones	yes no	
4. 17 + 7	____ ten ____ ones	yes no	
5. 32 + 6	____ tens ____ ones	yes no	
6. 59 + 5	____ tens ____ ones	yes no	
7. 13 + 9	____ ten ____ ones	yes no	
8. 31 + 8	____ tens ____ ones	yes no	
9. 25 + 6	____ tens ____ ones	yes no	
10. 62 + 7	____ tens ____ ones	yes no	

Problem Solving

Use Work Mat 6 and base-ten blocks to solve.

11. In April, it rained 18 days. It rained 6 days in May. How many days did it rain in April and May?

_____ days

12. John sold 17 bottles of water. Later, he sold 8 bottles of water. How many bottles of water did John sell in all?

_____ bottles of water

13. The runners ate 6 bananas before the race. They ate 14 bananas after the race. How many bananas did the runners eat in all?

_____ bananas

 Write Math   Explain how you can regroup 19 + 7 to find the sum.

_____

_____

_____

My Homework

Homework Helper Need help? connectED.mcgraw-hill.com

Find 13 + 8.

Step 1 Show each number.

Step 2 If there are 10 or more ones, regroup.

Step 3 Write the number of tens and ones.

tens	ones

13 + 8 = 21

Practice

Circle the ones to show regrouping ten ones as 1 ten. Draw your answer.

Helpful Hint
10 ones equals 1 ten.

1. 16 + 5 = _____

tens	ones

2. 18 + 4 = _____

tens	ones

Circle the ones to show regrouping ten ones as 1 ten. Draw your answer.

3. $14 + 8 =$ _____

tens	ones

4. $15 + 5 =$ _____

tens	ones

5. Max donated 16 dog collars to the dog shelter. Mia donated 6 leashes. How many items were donated?

_____ items

That collar is perfect for me!

Vocabulary Check

Circle the picture that matches the word.

6. regroup

Math at Home Using toothpicks have your child show you $25 + 5$. Ask how many tens there are in the answer.

Add to a Two-Digit Number

Lesson 3

ESSENTIAL QUESTION
How can I add two-digit numbers?

Explore and Explain Watch ▶ Tools

Look how tall I am Morgan!

tens	ones

_____ newspapers

Teacher Directions: Use base-ten blocks to show this addition story. Morgan and her friends are collecting newspapers to recycle. One group collects 7 newspapers. The other group collects 15 newspapers. How many newspapers have they collected in all? Write the number.

See and Show

Find 17 + 5.

Step 1
Add the ones.
7 + 5 = 12 ones

tens	ones
☐	
1	7
+	5

Step 2
Regroup if needed.
Write how many.

tens	ones
⬚	
1	7
+	5
	2

Step 3
Add the tens.
1 ten + 1 ten
= 2 tens

tens	ones
1	
1	7
+	5
2	2

Use Work Mat 6 and base-ten blocks to add.

1.

tens	ones
☐	
1	8
+	6

2.

tens	ones
☐	
4	3
+	3

3.
```
  ☐
  1  9
+    6
_____
```

4.
```
  ☐
  3  6
+    5
_____
```

5.
```
  ☐
  1  4
+    9
_____
```

Talk Math   How did you show that you regrouped?

Name

On My Own

Use Work Mat 6 and base-ten blocks to add.

6.
```
    □
   5 │ 6
 + │ 1
___│___
```

7.
```
    □
   7 │ 2
 + │ 8
___│___
```

8.
```
    □
   3 │ 8
 + │ 8
___│___
```

9.
```
    □
   2 │ 4
 + │ 4
___│___
```

10.
```
    □
   4 │ 3
 + │ 9
___│___
```

11.
```
    □
   1 │ 3
 + │ 7
___│___
```

12.
```
    □
   5 │ 1
 + │ 9
___│___
```

13.
```
    □
   1 │ 7
 + │ 6
___│___
```

14.
```
    □
   3 │ 8
 + │ 4
___│___
```

15.
```
   3 3
 +   7
_____
```

16.
```
   2 2
 +   9
_____
```

17.
```
   6 8
 +   3
_____
```

Problem Solving

18. Jaya has 25 stickers on her paper. She adds 4 more stickers. How many stickers does Jaya have now?

_____ stickers

19. Micah does 19 cartwheels. Then he does 8 more. How many cartwheels does Micah do altogether?

_____ cartwheels

20. 13 children are waiting for the bus. 5 more children get in line for the bus. How many children are waiting for the bus now?

_____ children

Write Math Explain how you can solve 14 + 7.

_____

_____

_____

Name ....

My Homework

Homework Helper eHelp   **Need help?** connectED.mcgraw-hill.com

Find 56 + 8.

Step 1 Add the ones.
Step 2 Regroup if needed.
Step 3 Add the tens.

So, 56 + 8 = 64.

tens	ones
[1]	
5	6
+	8
6	4

Practice

Add.

1.
```
    6 | 6
  + | 6
  ──┼──
```

2.
```
    8 | 4
  + | 3
  ──┼──
```

3.
```
    2 | 9
  + | 4
  ──┼──
```

4.
```
    3 | 7
  + | 9
  ──┼──
```

5.
```
    2 | 1
  + | 9
  ──┼──
```

6.
```
    4 | 5
  + | 6
  ──┼──
```

Add.

7.
$$\begin{array}{r} \square \\ 3\,|\,8 \\ +\quad|\,3 \\ \hline \end{array}$$

8.
$$\begin{array}{r} \square \\ 7\,|\,8 \\ +\quad|\,7 \\ \hline \end{array}$$

9.
$$\begin{array}{r} \square \\ 8\,|\,3 \\ +\quad|\,8 \\ \hline \end{array}$$

10. There are 24 students in Ms. Ito's class. 5 more students join her class. How many students are there now?

_____ students

Coach Harry is awesome!

11. Harry coaches 25 children on Wednesday. On Saturday, he coaches 9 more children. How many children does he coach in all?

_____ children

Test Practice

12. There are 54 people that arrive early for the Tigers' game. 9 people arrive late. How many people came to the game?

63 ○   52 ○   44 ○   20 ○

 Math at Home Ask your child to show you how to add 14 and 8. Then ask your child to show you how to add 27 and 2.

Name _____

Vocabulary Check

Complete each sentence.

sum   **add**   **regroup**   **addends**

1. You can _____ 10 ones into 1 ten.

2. The _____ is the answer to an addition problem.

3. You _____ by joining two numbers together.

4. The numbers you join when adding are

    called _____ .

Concept Check

Take apart an addend to solve.

5.   26 + 15

26 + _____ + _____

_____ + _____ = _____

So, 26 + 15 = _____ .

6.   18 + 54

18 + _____ + _____

_____ + _____ = _____

So, 18 + 54 = _____ .

Add.

7.
```
  4 │ 5
+   │ 9
────┼───
```

8.
```
  7 │ 3
+   │ 8
────┼───
```

9.
```
  2 │ 6
+   │ 3
────┼───
```

10.
```
  24
+  9
────
```

11.
```
  43
+  7
────
```

12.
```
  18
+  4
────
```

13. Liam's boy scout troop collected 18 bins of plastic bottles to recycle. Landon's troop collected 9 bins. How many bins did the boy scouts collect in all?

_____ bins

Test Practice

14. 31 people are in the store. 9 more people come into the store. How many people are in the store now?

| 40 | 31 | 22 | 9 |
| ○ | ○ | ○ | ○ |

Add Two-Digit Numbers

Explore and Explain

 Watch Tools

Flower sale!

Week 1	Week 2

_____ + _____ = _____

 Teacher Directions: Model this addition story with base-ten blocks. Mr. Kay's class is selling flowers. The first week they sold 45 flowers. The second week they sold 26 flowers. How many flowers did they sell in all? Write the number sentence.

See and Show

Find 18 + 25.

Step 1
Add the ones.
Regroup if needed.

tens	ones
☐ 1	
1	8
+ 2	5
	3

Step 2
Add the tens.

tens	ones
☐ 1	
1	8
+ 2	5
4	3

So, 18 + 25 = __43__

Use Work Mat 6 and base-ten blocks to add.

1.

tens	ones
☐	
4	7
+ 2	9

2.

tens	ones
☐	
1	2
+ 2	4

3.

tens	ones
☐	
8	2
+ 1	4

4.

tens	ones
☐	
1	5
+ 1	9

5.

tens	ones
☐	
2	4
+ 1	8

Talk Math What do you do first when you add
two-digit numbers?

Name _____

On My Own

Use Work Mat 6 and base-ten blocks to add.

6.

tens	ones
☐	
3	3
+ 1	8

7.

tens	ones
☐	
6	4
+ 2	6

8.

tens	ones
☐	
5	8
+ 1	2

9.
☐
$$\begin{array}{r} 3 \ | \ 6 \\ + 2 \ | \ 7 \\ \hline \end{array}$$

10.
☐
$$\begin{array}{r} 2 \ | \ 2 \\ + 1 \ | \ 0 \\ \hline \end{array}$$

11.
☐
$$\begin{array}{r} 5 \ | \ 9 \\ + 1 \ | \ 3 \\ \hline \end{array}$$

12.
☐
$$\begin{array}{r} 4 \ | \ 6 \\ + 2 \ | \ 6 \\ \hline \end{array}$$

13.
☐
$$\begin{array}{r} 2 \ | \ 8 \\ + 1 \ | \ 8 \\ \hline \end{array}$$

14.
☐
$$\begin{array}{r} 9 \\ + 3 \ | \ 2 \\ \hline \end{array}$$

15.
$$\begin{array}{r} 42 \\ + 42 \\ \hline \end{array}$$

16.
$$\begin{array}{r} 7 \\ + 27 \\ \hline \end{array}$$

17.
$$\begin{array}{r} 8 \\ + 11 \\ \hline \end{array}$$

18. Tia fed her turtle 30 times in June and 31 times in July. How many times did she feed her turtle?

_____ times

19. There are 25 chairs in one classroom. There are 28 chairs in another classroom. How many chairs are there in all?

_____ chairs

20. The pet store has 14 small lizards. There are also 16 large lizards. How many lizards are at the pet store in all?

_____ lizards

 How is adding 21 + 9 different than adding 21 + 19?

_____

_____

_____

Name _____

My Homework

Homework Helper Need help? connectED.mcgraw-hill.com

Find 25 + 17.

Step 1 Add the ones.
Regroup if needed.

Step 2 Add the tens.

So, 25 + 17 = 42.

tens	ones
1	
2	5
+ 1	7
4	2

Practice
Add.

1.

tens	ones
☐	
3	5
+ 2	5

2.

tens	ones
☐	
4	8
+ 1	8

3.

tens	ones
☐	
1	5
+ 5	9

4.
```
  ☐
  5 │ 6
+ 3 │ 5
────────
```

5.
```
  ☐
  2 │ 7
+ 2 │ 8
────────
```

6.
```
  ☐
  1 │ 4
+ 3 │ 3
────────
```

Add.

7.   53
    + 35

8.   23
    + 58

9.   83
    + 13

10.   62
     + 28

11.   53
     + 39

12.   49
     + 24

13. Brad's family picks 22 pounds of cherries. Seth's family picks 43 pounds. How many pounds of cherries did they pick in all?

_____ pounds

Test Practice

14. 46 people came to the museum early. 39 more people came late. How many people came to the museum in all?

86   85   39   7
○   ○   ○   ○

Math at Home Take two 2-digit numbers from your phone number and have your child add them. Example: 555-1234; 12 + 34 = 46

Name ....

Rewrite Two-Digit Addition

Explore and Explain

Emily's going to get a hit. Just watch!

$$11 + 8$$

tens	ones
+	

That's Emily.

_____ runs

 Teacher Directions: Emily's baseball team is playing a baseball game to raise money for a children's hospital. Emily's team scored 11 runs in the game. The other team scored 8 runs. How many runs were scored in all? Write the numbers in the place-value chart and add.

See and Show

You can rewrite a problem to add.

Find 35 + 26.

Step 1 Write one addend below the other addend. Line up the tens and the ones.

Step 2 Add. Regroup if necessary.

```
    1
   3 5
 + 2 6
 ------
   6 1
```

Rewrite the problem. Add.

1. 64 + 22

2. 26 + 65

3. 36 + 36

4. 73 + 19

5. 47 + 18

6. 56 + 37

Talk Math Why is it helpful to rewrite addition?

On My Own

Rewrite the problem. Add.

7. 26 + 36

+ _____

8. 17 + 19

+ _____

9. 74 + 16

+ _____

10. 18 + 63

+ _____

11. 64 + 27

+ _____

12. 73 + 18

+ _____

13. 56 + 23

+ _____

14. 37 + 39

+ _____

15. 28 + 33

+ _____

Problem Solving

Processes
&Practices

16. 38 hot dogs were sold during the baseball game. 49 hamburgers were sold. How many hot dogs and hamburgers were sold in all?

_____ hot dogs and hamburgers

17. 65 boys and 29 girls came to watch the baseball game. How many boys and girls came to watch the game?

All for one!

_____ boys and girls

HOT Problem There are 13 players on the Panthers baseball team, 15 players on the Eagles team, and 14 players on the Lions team. How many players are on all three teams? Explain.

_____

_____

_____

Name

My Homework

Homework Helper **Need help?** connectED.mcgraw-hill.com

Rewrite 36 + 49 to add.

Step 1 Write one addend below the other addend.

Step 2 Add. Regroup if necessary.

$$
\begin{array}{r}
1\\
3\ \ 6\\
+\ 4\ \ 9\\
\hline
8\ \ 5
\end{array}
$$

Helpful Hint
Line up the ones and the tens.

Practice

Rewrite the problem. Add.

1. 64 + 15

+

2. 26 + 57

+

3. 61 + 28

+

4. 37 + 47

+

5. 49 + 38

+

6. 34 + 18

+

Rewrite the problem. Add.

7. 72 + 17

+ _____

8. 19 + 18

+ _____

9. 29 + 35

+ _____

10. 43 girls ran the race. 38 boys ran the race. How many people ran the race in all?

Gosh, I hope they don't drop me!

Relay!

_____ people

Test Practice

11. Mark the answer that shows how to rewrite and solve 46 + 38.

```
  46
+ 38
  74
  ○
```

```
 316
  4̶6̶
- 38
   8
   ○
```

```
  1
  46
+ 38
  84
  ○
```

```
  1
  46
+ 38
  85
  ○
```

Math at Home Write 35 + 38 on a piece of paper. Have your child rewrite the problem and solve it.

Name

Add Three and Four Two-Digit Numbers

Explore and Explain
Watch Tools

CANNONBALL!

tens	ones

_____ cars

Teacher Directions: Miley's class is having a car wash. They washed 13 blue cars, 22 red cars, and 14 white cars. How many cars did they wash in all? Use base-ten blocks to show each number. Trace the blocks. Write the sum.

Online Content at **connectED.mcgraw-hill.com**

See and Show

You can add three and four two-digit numbers. Line up the ones and the tens. Add the ones first, then add the tens.

Helpful Hint
Remember, when you add the ones, look for a fact you know.

Add three numbers.

```
  3 6
  1 4        6 + 4 = 10
+ 2 4
-----
 74         10 + 4 = 14
```

Add four numbers.

```
  1 2
  3 5        4 + 4 = 8
  1 4
+ 2 4        8 + 5 = 13
-----
 85         13 + 2 = 15
```

Add.

1.
```
   17
   23
 + 14
```

2.
```
   22
   12
 + 15
```

3.
```
   31
   19
 + 25
```

4.
```
   18
   10
 + 32
```

5.
```
   12
   19
   21
 + 17
```

6.
```
   13
   22
   42
 + 16
```

7.
```
   11
   25
   33
 + 13
```

8.
```
   13
   42
    8
 + 36
```

Talk Math How is adding three two-digit numbers like adding two two-digit numbers?

Name

On My Own

Add.

Everyone stick together now!

9.
```
  25
  11
+ 15
```

10.
```
  51
  12
+ 32
```

11.
```
   2
  25
+ 42
```

12.
```
  13
  33
+ 45
```

13.
```
  22
  18
  32
+ 13
```

14.
```
  15
  31
  19
+ 20
```

15.
```
  31
  34
  14
+ 17
```

16.
```
  11
  15
  46
+ 15
```

17.
```
  34
  13
+ 13
```

18.
```
  16
  27
+ 36
```

19.
```
  61
  10
+ 19
```

20.
```
  32
  15
+ 38
```

21.
```
  12
  34
  14
+ 26
```

22.
```
  10
  43
  17
+ 20
```

23.
```
  11
  28
  24
+ 36
```

24.
```
  22
  36
  14
+ 23
```

25. Logan has 11 paper clips. Amanda has 31 paper clips. Nate has 26 paper clips. How many paper clips do they have altogether?

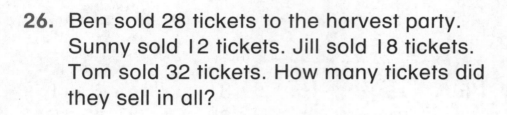

We'll need a lot of teamwork to untangle this mess!

_____ paper clips

26. Ben sold 28 tickets to the harvest party. Sunny sold 12 tickets. Jill sold 18 tickets. Tom sold 32 tickets. How many tickets did they sell in all?

_____ tickets

HOT Problem 14 boys play basketball. 12 boys play soccer. 16 boys play baseball. How many boys play basketball, soccer, or baseball?

Jayden solved the problem like this:
14 + 12 + 16 = 312 boys

Tell why he is wrong. Make it right. Explain.

_____

_____

_____

Name ....

Homework Helper **Need help?** connectED.mcgraw-hill.com

To add three or four two-digit numbers, line up the ones and the tens.

Add the ones, then add the tens.

```
   15
   16
 + 46
 ─────
   77
```

```
   18
   21
   46
 + 12
 ─────
   97
```

Helpful Hint
Look for a fact you know or doubles.

Practice

Add.

1.
```
   53
   27
 + 10
 ─────
```

2.
```
   52
   23
 + 18
 ─────
```

3.
```
   11
   19
 + 24
 ─────
```

4.
```
   26
   35
   24
 + 11
 ─────
```

5.
```
   23
   36
   16
 + 11
 ─────
```

6.
```
   23
   33
   13
 + 15
 ─────
```

Read and solve the problem.

7. 13 students play the bells. 16 students play the drums. 24 students play the recorder. 14 students play trumpet. How many students play instruments?

_____ students

8. Mr. Tan picks 24 apples. Mrs. Tan picks 35 apples. Their son picks 26 apples. How many apples do the Tans pick in all?

Take us to your toy drive earthling!

_____ apples

Test Practice

9. A class collects 20 toys on Monday, 15 toys on Tuesday and 10 toys on Wednesday. How many toys did they collect on Monday and Wednesday?

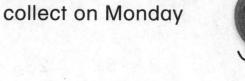

   10   20   30   35
  ○   ○   ○   ○

Math at Home Have your child explain how to add 28 + 12 + 35.

Problem Solving

STRATEGY: Make a Model

Tools

Mark and Dan were in a walk-a-thon to raise money for their basketball team. Mark walked 35 blocks. Dan walked 52 blocks. How many blocks did they walk in all?

Come on team. Let's walk!

1 Understand Underline what you know.
Circle what you need to find.

2 Plan How will I solve the problem?

3 Solve Make a model.

87 blocks

4 Check Is my answer reasonable? Explain.

Practice the Strategy

I can hear the ocean Jonny!

Sara, Mary, and Jonny are collecting shells at the beach. Sara finds 45 shells. Jonny finds 23 shells. Mary finds 15 shells. How many shells do they find?

1 Understand Underline what you know.
Circle what you need to find.

2 Plan How will I solve the problem?

3 Solve I will...

_____ shells

4 Check Is my answer reasonable? Explain.

Name _____

Apply the Strategy

1. Karenna ran 15 blocks Tuesday.
    She ran 20 blocks on Wednesday.
    How many blocks did she run in all?

    _____ blocks

2. Bradley got 33 action figures for his birthday.
    He had 12 already. How many will he have
    if he buys 15 more?

    _____ action figures

3. Nora made 16 points in her basketball game.
    Jenna made 24 points and Alicia made 22.
    No one else scored on their team.
    What was their final score?

    Great shooting **team!**

    _____ points

Review the Strategies

Choose a strategy
- Write a number sentence.
- Find a pattern.
- Make a model.

4. Melissa has 23 pairs of earrings. She buys 2 more pairs. Then she is given 6 more pairs for her birthday. How many pairs of earrings does she have in all?

_____ pairs

5. Jessica made muffins for the bake sale. She made 24 blueberry muffins and 10 strawberry muffins. She also made 12 banana muffins. How many muffins did she make in all?

Bake sale!

_____ muffins

6. Jacob and Luis collected glass bottles to recycle. The first day they collected 12 bottles. The second day they collected 15. The third day they collected 18. If this pattern continues, how many will they collect on the fourth day?

_____ bottles

Name

My Homework

eHelp

Karen and Jake are counting ribbons.
Jake counts 20 blue ribbons. Karen counts
16 red ribbons. How many ribbons did they
count in all?

First prize
in ribbon counting
goes to...

1ST PLACE

1 Understand Underline what you know.
Circle what you need to find.

2 Plan How will I solve the problem?

3 Solve Make a model.

$$20 + 16 = 36 \text{ ribbons}$$

4 Check Is my answer reasonable?

Underline what you know. Circle what you need to find. Make a model to solve.

1. Angela and Brittany count 8 windows in the first building, 5 in the next building, and 13 in the last building. How many windows did they count in all?

_____ windows

2. Eric found 14 pennies in the couch, 6 pennies in his mom's car, and 2 pennies under his bed. How many pennies did he find in all?

_____ pennies

3. The girls collected 57 bottle caps. The boys collected 42 caps. How many bottle caps did they collect in all?

_____ caps

Math at Home Ask your child to add 13 + 21 + 37. Have your child explain how he or she added the numbers.

208   Chapter 3 • Lesson 7

Fluency Practice

Add

1.   14
     + 3

2.   16
     + 6

3.   13
     + 8

4.   14
     + 9

5.   15
     + 6

6.   16
     + 3

7.   11
     + 9

8.   12
     + 2

9.   14
     + 0

10.   18
     + 5

11.   19
     + 1

12.   17
     + 2

Fluency Practice

Add

1. 13
   + 6

2. 16
   + 2

3. 13
   + 8

4. 26
   + 5

5. 13
   + 1

6. 13
   + 3

7. 21
   + 9

8. 16
   + 2

9. 11
   + 0

10. 13
    + 2

11. 13
    + 4

12. 12
    + 7

Together, we make beautiful music!

Name

Vocabulary Check

Draw lines to match.

1. **regroup**

2. **sum**

3. **add**

3 + 4 = 7

$$
\begin{array}{r}
\boxed{1} \\
34 \\
+\ 17 \\
\hline
51
\end{array}
$$

8 + 1 = 9
↑

Concept Check

Take apart an addend to solve.

4. 33 + 58

____ + ____ + 58

____ + ____ = ____

So, 33 + 58 = _____.

5. 17 + 46

17 + ____ + ____

____ + ____ = ____

So, 17 + 46 = _____.

Add.

6.

```
  □
  3 │ 4
+ 2 │ 7
──────
```

7.

```
  □
  4 │ 3
+ 2 │ 8
──────
```

8.

```
  □
  1 │ 5
+ 3 │ 7
──────
```

9.

```
  □
  1 │ 1
+ 1 │ 8
──────
```

10.

```
  □
  1 │ 5
+ 7 │ 3
──────
```

11.

```
  □
  2 │ 8
+ 6 │ 3
──────
```

12.

```
  □
    36
 + 36
──────
```

13.

```
  □
    26
 + 45
──────
```

14.

```
  □
    32
 + 16
──────
```

15.

```
    23
    17
 + 30
──────
```

16.

```
    52
    22
 + 20
──────
```

17.

```
    16
    12
 + 24
──────
```

18.

```
    16
    25
    44
 + 11
──────
```

19.

```
    57
    17
    13
 + 10
──────
```

20.

```
    21
    13
    13
 + 16
──────
```

Name _____

 Problem Solving

21. Kaylee sold 35 raffle tickets. Chloe sold 27 and Noah sold 33. How many raffle tickets did they sell in all?

_____ tickets

22. Peyton has 13 marbles. She finds 11 more marbles. How many marbles does Peyton have in all?

_____ marbles

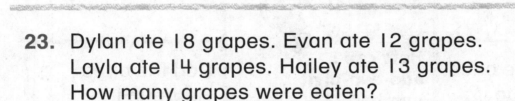

We're on a roll!

23. Dylan ate 18 grapes. Evan ate 12 grapes. Layla ate 14 grapes. Hailey ate 13 grapes. How many grapes were eaten?

_____ grapes

Test Practice

24. The red team has 20 boys, the blue team has 19 boys, the green team has 19 boys, and the yellow team has 20 boys. How many boys are there in all?

80 ○   78 ○   59 ○   58 ○

Show the ways you can add to solve problems.

Add the ones. Regroup
if needed.

$$\begin{array}{r} 17 \\ + \ 8 \\ \hline \end{array}$$

Add the ones. Regroup
if needed. Add the tens.

$$\begin{array}{r} 24 \\ + 17 \\ \hline \end{array}$$

**ESSENTIAL
QUESTION**

**How can I
add two-digit
numbers?**

Add the ones.
Regroup if
needed. Add
the tens.

$$\begin{array}{r} 18 \\ 12 \\ + 14 \\ \hline \end{array}$$

Add the ones.
Regroup if
needed. Add
the tens.

$$\begin{array}{r} 24 \\ 35 \\ 10 \\ + 15 \\ \hline \end{array}$$

Don't stop
until you reach
the top!

Go team! Go!

Chapter 4 Subtract Two-Digit Numbers

Let's Explore the Seasons!

Watch a video!

Watch ▶

Chapter 4 Project

Teach a Review Mathematics Lesson

1. Work together with your group to write a review lesson from this chapter which your group will need to present to the class. Each group should present their lesson for 5–8 minutes.

2. Create an outline or lesson plan for the lesson you plan to teach.

3. Your group should plan to use manipulatives, visuals, posters, props, or technology to teach the lesson.

4. Make sure to plan your lesson and practice it several times before presenting it to the class.

Date: _____

Topic or Concept: _____

Materials: _____

Lesson Plan

_____

_____

_____

_____

_____

_____

Name _____

Am I Ready?

Write how many.

1. How many tens are in 30? _____

2. How many tens are in 70? _____

3. How many tens are in 90? _____

Subtract.

4. 9 − 6 = _____

5. 8 − 2 = _____

6. 7 − 2 = _____

7. 6 − 5 = _____

8. 4 − 3 = _____

9. 5 − 3 = _____

Write a number sentence.

10. Maggie has 10 pretzels in her lunch. Lamont eats 4 of Maggie's pretzels. How many pretzels are left?

_____ ◯ _____ ◯ _____ pretzels

How Did I Do?

Shade the boxes to show the problems you answered correctly.

1	2	3	4	5	6	7	8	9	10

My Math Words

Vocab
abc

Review Vocabulary

| difference | fact family | sum |

**Complete the addition and subtraction sentences.
Complete each sentence.**

$18 - 6 =$ ☐

$18 - 12 =$ ☐

$12 + 6 =$ ◯

$6 + 12 =$ ◯

The boxed numbers are called the _____.

The circled numbers are called the _____.

The numbers 6, 12, and 18 are called a _____.

My Vocabulary Cards

Vocab
abc

Processes
&Practices

Directions:
Ideas for Use
- Have students use blank cards to write subtraction facts they know.
- Ask students to use the blank cards to write their own vocabulary cards.

My Foldable

51

42

93

7

3

10

66

33

99

Name

Two-Digit Fact Families

Explore and Explain Watch

I love snow days!

$$\_\_\_\_ + \_\_\_\_ = \_\_\_\_ \qquad \_\_\_\_ - \_\_\_\_ = \_\_\_\_$$

$$\_\_\_\_ + \_\_\_\_ = \_\_\_\_ \qquad \_\_\_\_ - \_\_\_\_ = \_\_\_\_$$

 Teacher Directions: Use base-ten blocks to solve. Lindsay has 16 snowballs. She throws 7 of them. How many are left? Write the three numbers in the fact family. Then write the related number sentences.

See and Show

The numbers in this addition and subtraction fact family are 18, 15, and 3.

$15 + 3 = \underline{18}$ $18 - 15 = \underline{3}$

$3 + 15 = \underline{18}$ $18 - 3 = \underline{15}$

Complete each fact family.

1.

$12 + 12 = \underline{\hspace{1cm}}$ $24 - 12 = \underline{\hspace{1cm}}$

2.

$11 + 6 = \underline{\hspace{1cm}}$ $17 - 11 = \underline{\hspace{1cm}}$

$6 + 11 = \underline{\hspace{1cm}}$ $17 - 6 = \underline{\hspace{1cm}}$

3.

$16 + 4 = \underline{\hspace{1.5cm}}$ $20 - 4 = \underline{\hspace{1.5cm}}$

$\underline{\hspace{1cm}} + \underline{\hspace{1cm}} = \underline{\hspace{1cm}}$ $\underline{\hspace{1cm}} - \underline{\hspace{1cm}} = \underline{\hspace{1cm}}$

Talk Math   How are these exercises like fact families that you have learned earlier?

....

On My Own

Complete each fact family.

4.

____ + ____ = ____   ____ − ____ = ____

____ + ____ = ____   ____ − ____ = ____

5.

____ + ____ = ____   ____ − ____ = ____

____ + ____ = ____   ____ − ____ = ____

6.

____ + ____ = ____   ____ − ____ = ____

____ + ____ = ____   ____ − ____ = ____

7.

____ + ____ = ____   ____ − ____ = ____

____ + ____ = ____   ____ − ____ = ____

8.

____ + ____ = ____   ____ − ____ = ____

____ + ____ = ____   ____ − ____ = ____

9. Lizzie has 12 necklaces without beads. She also has 22 necklaces that have beads on them. How many necklaces does Lizzie have? Write the fact family that can be made from these numbers.

_____ necklaces

____ + ____ = ____   ____ − ____ = ____

____ + ____ = ____   ____ − ____ = ____

10. How many parking spaces are open in the parking lot?

_____ parking spaces

26 Parking Spaces

14 Filled _____ Open

HOT Problem Sophia wrote this fact family. Tell why Sophia is wrong. Make it right.

$$76 + 12 = 88 \qquad 76 - 12 = 88$$
$$12 + 76 = 88 \qquad 12 - 76 = 88$$

_____

_____

_____

Name _____

Homework Helper Need help? connectED.mcgraw-hill.com

The numbers 15, 11, and 4 make up the numbers in a fact family.

$$11 + 4 = 15 \qquad 15 - 11 = 4$$
$$4 + 11 = 15 \qquad 15 - 4 = 11$$

Practice

Complete each fact family.

1.

____ + ____ = ____   ____ − ____ = ____

____ + ____ = ____   ____ − ____ = ____

2.

____ + ____ = ____   ____ − ____ = ____

____ + ____ = ____   ____ − ____ = ____

3.

____ + ____ = ____   ____ − ____ = ____

____ + ____ = ____   ____ − ____ = ____

Complete the fact family.

4.

26

13   13

_____ + _____ = _____   _____ − _____ = _____

Use what you know about fact families to solve.

5. Marcus counts 24 children playing. 11 of the children are boys. How many girls are playing?

Catch me if you can!

_____ girls

6. There are 49 second graders. 26 of them are in Miss Johnson's class. How many students are in Mrs. Stewart's class?

_____ students

Test Practice

7. Which number sentence does not belong in the fact family for the numbers 7, 18, and 25?

$18 + 7 = 25$
○

$18 − 25 = 7$
○

$7 + 18 = 25$
○

$25 − 7 = 18$
○

  Math at Home Have your child write the fact family for the numbers 22, 13, and 35.

Take Apart Tens to Subtract

Explore and Explain

Watch · Tools

Lesson 2

ESSENTIAL QUESTION
How can I subtract
two-digit numbers?

Leaf pile!

Leaf pile!

22 = _____ tens _____ ones

56 − _____ = _____

_____ − _____ = _____

_____ leaves

Teacher Directions: There are 56 leaves on a tree. 22 leaves fall off of the tree. How many leaves are left? Use base-ten blocks to show 56 leaves. Take apart 22 into tens and ones. Subtract the tens from 56. Then subtract the ones from that difference.

See and Show

You can take apart numbers to subtract.

To find 35 − 13, take apart a number to make a ten.
Then subtract.

35 − 13

10   3

35 − 10 = 25

25 − 3 = 22

So, 35 − 13 = 22.

Helpful Hint
Take apart 13 as
10 and 3.

**Take apart a number to make a ten.
Then subtract.**

1. 64 − 12

____ ____

64 − ____ = ____

____ − ____ = ____

So, 64 − 12 = ____.

2. 45 − 24

____ ____

45 − ____ = ____

____ − ____ = ____

So, 45 − 24 = ____.

Talk Math Explain how you decide what to subtract
when working these problems.

On My Own

**Take apart a number to make a ten.
Then subtract.**

3.   47 − 26

____   ____

47 − ____ = ____

____ − ____ = ____

So, 47 − 26 = ____.

4.   77 − 43

____   ____

77 − ____ = ____

____ − ____ = ____

So, 77 − 43 = ____.

5.   85 − 32

____   ____

85 − ____ = ____

____ − ____ = ____

So, 85 − 32 = ____.

6.   75 − 55

____   ____

75 − ____ = ____

____ − ____ = ____

So, 75 − 55 = ____.

![Real World] **Problem Solving**

Processes & Practices

7. Cara has 28 sticks of sidewalk chalk. She lets her friend borrow 11 of them. How many sticks of chalk does Cara have now?

_____ sticks of chalk

8. Shawna had 37 snowballs. She threw 23 of them. How many snowballs does Shawna have left?

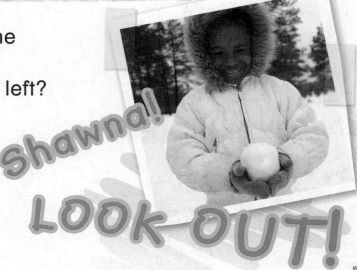

_____ snowballs

HOT Problem There are 22 students in Ms. Marshall's class. 8 students like science, 2 like art, and 3 like music. The rest of the students like math. How many students like math?

_____

_____

_____

Name _____

My Homework

Homework Helper

Need help? connectED.mcgraw-hill.com

To find $48 - 15$, take apart a number to make a ten.
Then subtract.

$$48 - 15$$

$$10 \qquad 5$$

$$48 - 10 = 38$$

$$38 - 5 = 33$$

So, $48 - 15 = 33$.

> **Helpful Hint**
> Take apart 15 as
> 10 and 5.

Practice

Take apart a number to make a ten. Then subtract.

1. $54 - 13$

54 _____ − _____ = _____

_____ − _____ = _____

So, $54 - 13 = $ _____.

2. $35 - 14$

35 _____ − _____ = _____

_____ − _____ = _____

So, $35 - 14 = $ _____.

3. There are 87 insects in a jar. Someone leaves the lid off and some crawl away. There are 54 insects left in the jar. How many crawled away?

_____ insects

4. Emily has 25 onions and 14 peppers that she grew in her garden. How many more onions does she have than peppers?

Come on out, Caleb!

_____ onions

5. There were 25 days of rain in April. It rained 16 days in May. If there are 61 days total in April and May, how many days did it not rain?

_____ days

Test Practice

6. How would you take apart 14 to solve 28 – 14?

   7 and 7 ◯   12 and 2 ◯

   10 and 4 ◯   20 and 8 ◯

Math at Home Have your child take apart 16 to make a ten to find 87 – 16.

Regroup a Ten as Ones

Lesson 3

ESSENTIAL QUESTION
How can I subtract
two-digit numbers?

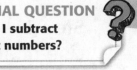

Explore and Explain

Watch ▶ Tools

Can you say
"She sells sea shells
by the sea shore."
really fast?

tens	ones

_____ sea shells

Teacher Directions: Adam has 23 sea shells. Use base-ten blocks to show the
sea shells. He gives 5 of them away. Regroup and take away that many blocks.
How many sea shells are left? Draw the base-ten blocks that are left in the chart.
Write how many sea shells are left.

See and Show

Find 24 − 8.

Step 1
Use base-ten blocks to show 24.

tens	ones

2 tens 4 ones

Step 2
Subtract the ones. There are not enough ones to subtract. Regroup.

tens	ones

I ten 14 ones

Step 3
Subtract 8 ones.

tens	ones

I ten 6 ones

$24 - 8 = \underline{16}$

Use Work Mat 6 and base-ten blocks to subtract.

	Subtract the ones. Do you need to regroup?	Write the difference.
1. 31 − 4	no yes	$31 - 4 =$ _____
2. 27 − 5	no yes	$27 - 5 =$ _____

 How do you know when you need to regroup? Explain.

Name _____

On My Own

Use Work Mat 6 and base-ten blocks to subtract.

	Subtract the ones. Do you need to regroup?	Write the difference.
3. 46 − 2	no yes	46 − 2 = _____
4. 35 − 9	no yes	35 − 9 = _____
5. 56 − 5	no yes	56 − 5 = _____
6. 22 − 6	no yes	22 − 6 = _____
7. 30 − 4	no yes	30 − 4 = _____
8. 43 − 3	no yes	43 − 3 = _____
9. 76 − 8	no yes	76 − 8 = _____
10. 67 − 8	no yes	67 − 8 = _____

Problem Solving

Use Work Mat 6 and base-ten blocks to subtract.

11. Logan draws 23 red flowers. He draws 14 yellow flowers. How many more red flowers than yellow flowers does he draw?

_____ red flowers

12. There are 25 flower bulbs in a bag. Karen gives 7 bulbs to Ava. How many bulbs are left?

I may not be pretty, but Mom says I'm a late bloomer.

_____ bulbs

13. Candice has played in 8 soccer games. There are 14 games in all. How many games does she have left?

_____ games

Write Math   Liam subtracts 5 from 23. Is the difference less than 20 or greater than 20? Explain.

_____

_____

_____

Name _____

My Homework

Homework Helper **Need help?** connectED.mcgraw-hill.com

Find 35 − 7.

Step 1
Show 35.

tens	ones

3 tens 5 ones

Step 2
Subtract the ones. There are not enough ones to subtract. Regroup.

tens	ones

2 tens 15 ones

Step 3
Subtract 7 ones.

tens	ones

2 tens 8 ones
35 − 7 = 28

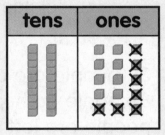

Helpful Hint
If there are not enough ones to subtract, regroup 1 ten as 10 ones.

Practice

Regroup the tens. Draw the ones. Cross out cubes to show subtraction. Write the answer.

1. 26 − 9 = _____

tens	ones

2. 23 − 5 = _____

tens	ones

Regroup the tens. Draw the ones. Cross out cubes to show subtraction. Write the answer.

3. 21 − 7 = _____

tens	ones

4. 30 − 6 = _____

tens	ones

5. There are 42 icicles. 9 of them melt. How many icicles are left?

_____ icicles

tens	ones

6. 31 people are at the skating rink. 8 people go home. How many people are left at the skating rink?

_____ people

tens	ones

Test Practice

7. 23 − 6 = ____

29 ○   17 ○   13 ○   10 ○

tens	ones

Math at Home Have your child use pennies to show you how to subtract 6 from 25.

Name _____

Subtract From a Two-Digit Number

Explore and Explain

 Watch Tools

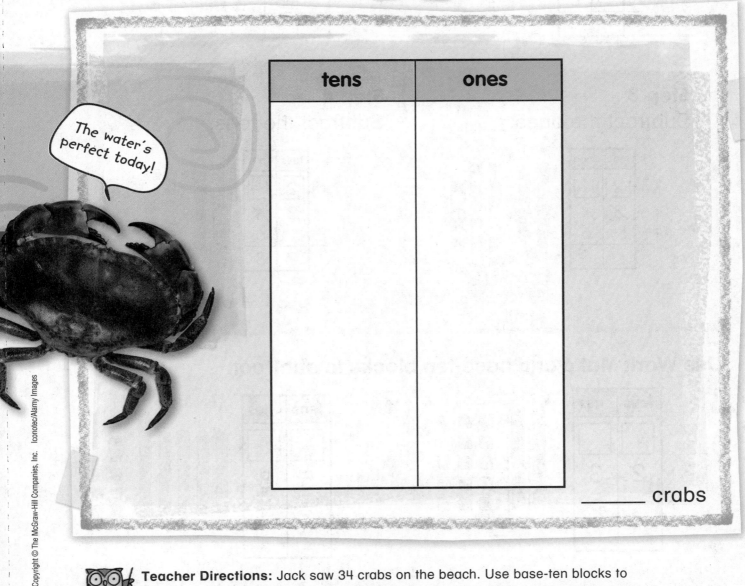

The water's perfect today!

tens	ones

_____ crabs

 Teacher Directions: Jack saw 34 crabs on the beach. Use base-ten blocks to show that number. 8 crabs ran into the water. Regroup and take away that many blocks. Draw the blocks that are left. Write how many crabs were left on the beach.

Online Content at ⟋ **connectED.mcgraw-hill.com**   Chapter 4 • Lesson 4   241

See and Show

Helpful Hint
Cross out each number as you regroup it.

Processes & Practices

Find 34 − 6.

Step 1
Show 34.
Can you subtract 6 ones?

Step 2
Regroup 1 ten as 10 ones.
4 ones + 10 ones = 14 ones.

Step 3
Subtract the ones.

Step 4
Subtract the tens.

34 − 6 = _28_

Use Work Mat 6 and base-ten blocks to subtract.

1.

2.

Talk Math How do you regroup 1 ten?

Name _____

On My Own

Use Work Mat 6 and base-ten blocks to subtract.

3.

tens	ones
3	4
−	7

4.

tens	ones
7	3
−	3

5.

tens	ones
5	9
−	4

6.

$$
\begin{array}{c|c}
\square & \square \\
8 & 8 \\
- & 2 \\
\hline
\end{array}
$$

7.

$$
\begin{array}{c|c}
\square & \square \\
1 & 5 \\
- & 9 \\
\hline
\end{array}
$$

8.

$$
\begin{array}{c|c}
\square & \square \\
3 & 1 \\
- & 6 \\
\hline
\end{array}
$$

9.

$$
\begin{array}{c|c}
\square & \square \\
5 & 8 \\
- & 9 \\
\hline
\end{array}
$$

10.

$$
\begin{array}{c|c}
\square & \square \\
6 & 6 \\
- & 7 \\
\hline
\end{array}
$$

11.

$$
\begin{array}{c|c}
\square & \square \\
9 & 5 \\
- & 5 \\
\hline
\end{array}
$$

12.
$$
\begin{array}{r}
14 \\
- \ 7 \\
\hline
\end{array}
$$

13.
$$
\begin{array}{r}
28 \\
- \ 9 \\
\hline
\end{array}
$$

14.
$$
\begin{array}{r}
34 \\
- \ 6 \\
\hline
\end{array}
$$

15. Tina puts 12 coins in a machine.
The machine keeps 4 of the coins.
How many coins does she get back?

_____ coins

16. There are 38 apples in the cafeteria. Students buy 9 apples for their lunches. How many apples are left?

_____ apples

17. A store has 76 shirts on display. The store sells 8 of the shirts. How many shirts are left?

_____ shirts

HOT Problem There are 23 boys and 33 girls at a beach. 7 boys and 4 girls leave the beach, how many boys and girls are left altogether? Explain.

_____

_____

_____

_____

Name
_____

My Homework

Homework Helper Need help? connectED.mcgraw-hill.com

Find 38 − 9.

	tens	ones
	2	18
	3̶	8̶
−		9
	2	9

Step 1   Can you subtract ones?

Step 2   Regroup 1 ten as 10 ones.

Step 3   Subtract the ones.

Step 4   Subtract the tens.

Practice

Subtract.

1.

tens	ones
2	6
−	9

2.

tens	ones
4	7
−	6

3.

tens	ones
6	2
−	5

4.

5.

6.

	4	9
−		6

Subtract.

7.
$$44$$
$$-\ 4$$

8.
$$51$$
$$-\ 8$$

9.
$$22$$
$$-\ 3$$

10. There are 27 children playing outside.
7 children go inside. How many are left?

_____ children

11. There are 55 mice in the barn. A cat chases
9 of them away. How many mice are left?

_____ mice

I like
barns!

Test Practice

12. Find the difference.

$$43 - 9 = \_\_\_\_$$

52 ⚪ 44 ⚪ 33 ⚪ 34 ⚪

I do
too!

Math at Home Ask your child to show you how to subtract 8 from 27
by regrouping.

Name _____

Vocabulary Check

Complete the sentence.

fact family   regroup

1. The numbers 3, 5, and 8 make up a _____.

Concept Check

Complete the fact family.

2.

19

7   12

___ + ___ = ___   ___ − ___ = ___

___ + ___ = ___   ___ − ___ = ___

Take apart a number to make a ten. Then subtract.

3. 83 − 52

∧

_____ _____

83 − _____ = _____

_____ − _____ = _____

So, 83 − 52 = _____.

4. 67 − 45

∧

_____ _____

67 − _____ = _____

_____ − _____ = _____

So, 67 − 45 = _____.

Subtract.

5.

tens	ones
☐	☐
6	2
−	8

6.

tens	ones
☐	☐
4	3
−	7

7.

tens	ones
☐	☐
5	5
−	4

8.

```
  ☐ | ☐
  2 | 5
−   | 6
```

9.

```
  ☐ | ☐
  5 | 4
−   | 8
```

10.

```
  ☐ | ☐
  7 | 3
−   | 6
```

11. There are 35 robins in a park. 14 of them fly away. How many robins are left in the park?

_____ robins

Test Practice

12. Yesterday, there were 77 tulips blooming in the garden. Today, 8 of them have wilted. How many tulips are still blooming?

69 ○   78 ○   79 ○   85 ○

Subtract Two-Digit Numbers

Lesson 5

ESSENTIAL QUESTION
How can I subtract two-digit numbers?

Explore and Explain

Watch | Tools

 I can't see anything with these goggles.

Pool Party!

tens	ones

_____ people

 Teacher Directions: There are 42 people swimming at the pool. Use base-ten blocks to show those people. 13 people go home. Take away 13 base-ten blocks. Regroup if needed. Draw the base-ten blocks that are left in the chart. Write how many people are left. How many are left at the pool?

See and Show

Processes & Practices

Find 52 − 17.

Step 1
Show 52.
Can you subtract 7 ones?

Step 2
Regroup 1 ten as 10 ones.
2 ones + 10 ones = 12 ones.

Step 3
Subtract the ones.

Step 4
Then subtract the tens.

52 − 17 = _____ 35

Use Work Mat 6 and base-ten blocks to subtract.

1.

2.

Talk Math How is subtracting 41 − 16 different than
41 − 6?

250   Chapter 4 • Lesson 5

Name

On My Own

Use Work Mat 6 and base-ten blocks to subtract.

3.

tens	ones
☐	☐
2	6
− 1	9

4.

tens	ones
☐	☐
7	4
− 1	8

5.

tens	ones
☐	☐
4	6
− 2	3

6.

☐	☐
7	5
− 1	6

7.

☐	☐
5	4
− 2	8

8.

☐	☐
7	2
− 3	5

9.

☐	☐
5	0
− 2	4

10.

☐	☐
8	2
− 3	7

11.

☐	☐
3	9
− 2	9

12.   85
      − 46

13.   41
      − 15

14.   37
      − 11

15. 35 people are in line for the diving board at the pool. 16 people jump off the diving board. How many people are still in line?

_____ people

16. Jason bought a pool pass for 45 trips to the pool. He has gone to the pool 28 times. How many more times can Jason go to the pool with his pass?

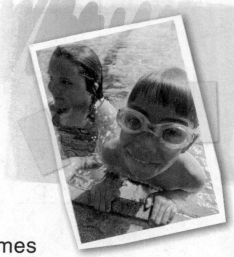

_____ times

HOT Problem Ella, May, and Sue each pour 24 cups of lemonade for their lemonade stand. They sell 23 of the cups of lemonade to their friends. How many cups are left to sell? Explain.

_____

_____

_____

Name _____

My Homework

Homework Helper   Need help? connectED.mcgraw-hill.com

Find 48 − 19.

Step 1 Can you subtract ones?

Step 2 Regroup 1 ten as 10 ones.

Step 3 Subtract the ones.

Step 4 Subtract the tens.

tens	ones
3̶4̶	1̶8̶
− 1	9
2	9

Practice

Subtract.

1.

tens	ones
□	□
6	3
− 2	5

2.

tens	ones
□	□
4	7
− 1	7

3.

tens	ones
□	□
2	4
− 1	6

4.

□	□
6	8
− 3	7

5.

□	□
5	3
− 2	4

6.

□	□
5	8
− 2	9

Subtract.

7.	54 − 45	8.	27 − 9	9.	63 − 11
10.	91 − 59	11.	35 − 26	12.	87 − 42

13. 83 people were at the water park. 29 people left. How many people were left at the water park?

Which way to the water?

_____ people

Test Practice

14. 25 people brought dogs to the picnic. 18 of them were large dogs. How many small dogs were at the picnic?

    6   7   13   17

    ○   ○   ○   ○

Math at Home Ask your child to show you how to subtract 24 from 41.

Name ....

Rewrite Two-Digit Subtraction

Explore and Explain Tools

Spring has sprung!

$$54 - 36$$

tens	ones
−	

_____ flowers

Teacher Directions: 54 flowers were planted in the garden. Write this number first. 36 flowers have bloomed. Write this number below. How many flowers still need to bloom? Rewrite the problem to subtract.

See and Show

You can rewrite a problem to subtract.
Find 83 − 28.

Step 1 Rewrite.

Step 2 Subtract.

Helpful Hint
Write the greater number on top. Write the other number below it.

Line up the ones and the tens.

Rewrite the problem. Subtract.

1. 38 − 19

2. 42 − 27

3. 53 − 38

4. 82 − 46

5. 74 − 38

6. 60 − 48

Talk Math Why is it helpful to rewrite subtraction?

Name _____

On My Own

Rewrite the problems. Subtract.

7. 62 − 38

8. 74 − 39

9. 57 − 39

10. 88 − 49

11. 35 − 25

12. 62 − 38

13. 67 − 19

14. 83 − 39

15. 90 − 36

Problem Solving

16. There are 64 flowers growing in Kyla's garden. She picks 18 flowers. How many flowers are left in the garden?

_____ flowers

17. Tyler kept track of the weather for 73 days. It was sunny 59 days. How many days were not sunny?

_____ days

18. Jackie has 32 houses on her street. She counted 17 that are brick. How many houses are not brick?

_____ houses

HOT Problem There are 91 children. 32 children play baseball. 21 children play soccer. How many children do not play baseball or soccer? Explain.

_____

_____

_____

Name ....

My Homework

Homework Helper Need help? connectED.mcgraw-hill.com

Rewrite to subtract 82 − 49.

```
   7   12
   8 | 2
 − 4 | 9
 ─────────
   3 | 3
```

> **Helpful Hint**
> When you subtract, always put the larger number on top.

Practice

Rewrite the problems. Subtract.

1. 74 − 25

2. 60 − 37

3. 86 − 48

4. 45 − 28

5. 84 − 38

6. 37 − 18

Rewrite the problems. Subtract.

7. 72 − 37

−

8. 27 − 18

−

9. 68 − 39

−

10. 45 birds are flying. 17 birds land on the water. How many birds are still flying?

_____ birds

I would never swim anywhere else!

SWIM AT BEACH ONLY

Test Practice

11. Mark the answer that shows how to rewrite and solve 83 − 46.

7 13
8̶3̶
−46
36
○

7 13
8̶3̶
−46
37
○

3 16
4̶6̶
−83
37
○

3 16
4̶6̶
−83
43
○

Math at Home Together with your child think of a 2-digit subtraction problem in your lives. Have them solve it.

Name

Check Subtraction

Lesson 7

ESSENTIAL QUESTION
How can I subtract
two-digit numbers?

Explore and Explain

Show your work.

☐
− ☐
─────
☐

Check your work.

☐
+ ☐
─────
☐

 Teacher Directions: There are 38 vegetables in the garden. Nancy picks 12
vegetables. How many vegetables are left? Show your work. Now check your work.
Write the number left in the blue box. Write the number of vegetables picked in the
yellow box. Add. What do you notice about the sum?

Online Content at ⌁ **connectED.mcgraw-hill.com**   Chapter 4 • Lesson 7   261

Copyright © The McGraw-Hill Companies, Inc. Ingram Publishing/Alamy Images

See and Show

You can check the answer to a subtraction problem. Check $25 - 10 = 15$.

Helpful Hint
Add the number you subtracted to the difference. The sum should match the number you subtracted from.

Subtract

$$\begin{array}{r} 25 \\ - \ 10 \\ \hline 15 \end{array}$$

Add these numbers to check.

Check by Adding

$$\begin{array}{r} 15 \\ + \ 10 \\ \hline 25 \end{array}$$

If this is the number you subtracted from, your answer is correct.

Subtract. Check by adding.

1. $\begin{array}{r} 36 \\ - \ 20 \\ \hline \end{array}$ $+ \ \underline{\quad}$

2. $\begin{array}{r} 50 \\ - \ 12 \\ \hline \end{array}$ $+ \ \underline{\quad}$

3. $\begin{array}{r} 21 \\ - \ 3 \\ \hline \end{array}$ $+ \ \underline{\quad}$

4. $\begin{array}{r} 32 \\ - \ 14 \\ \hline \end{array}$ $+ \ \underline{\quad}$

Talk Math Why does addition work as a check for subtraction?

Name _____

Helpful Hint
To check subtraction, add the number you subtracted and the difference.

On My Own

Subtract. Check by adding.

5.   57
    − 30   + ___

6.   60
    − 10   + ___

7.   42
    − 5   + ___

8.   75
    − 41   + ___

9.   67
    − 32   + ___

10.   40
     − 26   + ___

11.   76
     − 28   + ___

12.   56
     − 28   + ___

13. Ima went to the apple orchard with her class. The class picked 48 apples. They ate 17 of them. How many apples are left?

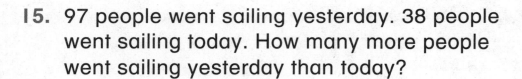

_____ apples

14. The second grade classes are planting seeds. Roy's class plants 81 seeds. Jessica's class plants 66 seeds. How many more seeds does Roy's class plant than Jessica's class?

_____ seeds

15. 97 people went sailing yesterday. 38 people went sailing today. How many more people went sailing yesterday than today?

_____ people

 Write Math   How do you check a subtraction problem?

_____

_____

_____

Name _____

My Homework

Homework Helper Need help? connectED.mcgraw-hill.com

You can check the answer to a subtraction problem.

Check $37 - 15 = 22$.

Subtract		**Check by Adding**	
37	Add these	22	If this is the
− 15	numbers to	+ 15	number you
22	check.	37	subtracted from, you are correct.

Practice

Subtract. Check by adding.

1. 67
   − 48   + ___

2. 52
   − 36   + ___

3. 80
   − 68   + ___

4. 91
   − 45   + ___

Solve. Check by adding.

Ice is nice!

5. There are 46 girls skating.
There are 67 boys skating.
How many more boys than
girls are skating?

_____ boys

6. Molly checks out 21 books from the library.
She returns 12. How many books does
Molly still have?

_____ books

7. Reid kicked the football 20 times. He missed
a field goal 6 times. How many times did he
make a field goal?

_____ times

Test Practice

8. Mrs. Levine is 83 years old. Mrs. Smith is 67 years old.
How many years older is Mrs. Levine than Mrs. Smith?

16 years   20 years   26 years   17 years
   ○   ○   ○   ○

Math at Home Ask your child to find 43 − 16. Then have your child show you how
to check the answer.

Name

Problem Solving
STRATEGY: Write a Number Sentence

There are 24 bees on a flower. 15 bees fly away. How many bees are now on the flower?

1 Understand Underline what you know.
Circle what you need to find.

2 Plan How will I solve the problem?

3 Solve Write a number sentence.

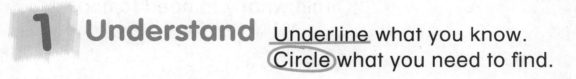

$$24 \ominus 15 = \underline{9} \text{ bees}$$

4 Check Is my answer reasonable? Explain.

Practice the Strategy

84 children are at summer camp. 29 children leave to come home early. How many children are still at summer camp?

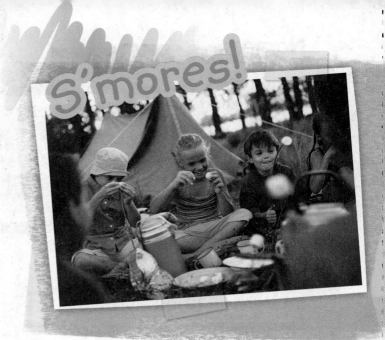

S'mores!

1 Understand Underline what you know.
Circle what you need to find.

_____

2 Plan How will I solve the problem?

_____

3 Solve I will...

_____ ◯ _____ ◯ _____

_____

4 Check Is my answer reasonable? Explain.

Processes & Practices

Apply the Strategy

Write a number sentence to solve.

1. There are 25 ants in an ant hill.
13 ants leave. How many ants
are there now?

_____ ◯ _____ ◯ _____ ants

2. There are 18 lions in the yard at the zoo.
7 lions run into the lion house.
How many lions are left in the yard?

_____ ◯ _____ ◯ _____ lions

3. Jay planted 12 daisies. His sister, Sarah,
planted 29 daisies. How many more
daisies did Sarah plant than Jay?

_____ ◯ _____ ◯ _____ daisies

Review the Strategies

Choose a strategy
- Find a pattern.
- Write a number sentence.
- Act it out.

4. Brenda collects 39 leaves. Janet collects 45 leaves. How many more leaves did Janet collect than Brenda?

_____ leaves

5. Anna is picking vegetables. She picks 12 tomatoes, 15 peppers, and 5 cucumbers. How many vegetables does she pick?

_____vegetables

6. Scott has 63 blueberries to put in a pie. He eats 21 of the blueberries. How many blueberries are in the pie?

_____ blueberries

7. Trey's dog likes dog treats. There are 60 treats in a box. He is allowed 2 treats a day. How many treats will he have left after 10 days?

_____ treats

Name

My Homework

Tara and her dad are baking bread
for a bake sale. They make 34 loaves
of bread. 11 loaves were sold.
How many loaves are left?

No loafing
at this
bake sale!

1 Understand Underline what you know.
Circle what you need to find.

2 Plan How will I solve the problem?

3 Solve I will write a number sentence.

34 − 11 = 23 loaves

4 Check Is my answer reasonable?

Practice

Write a number sentence to solve.

1. Emma has 31 ribbons for her hair. She loses 12 ribbons. How many does she have left?

_____ ◯ _____ ◯ _____ ribbons

Living on a pond isn't what it's quacked up to be.

2. Mason's soccer team has 24 games this season. He has already played 18 games. How many games are left?

_____ ◯ _____ ◯ _____ games

3. There are 18 ducks in the pond. 11 ducks get out of the pond. How many are still in the pond?

_____ ◯ _____ ◯ _____ ducks

Test Practice

4. Karen took a walk on the beach and counted 25 sandcastles. On the way back she counted again but some were gone. There were only 16 left. How many had disappeared while she walked?

11 ◯   9 ◯   19 ◯   41 ◯

Math at Home Ask your child to subtract 52 − 17. Then have your child show you how to check the answer.

Two-Step Word Problems

Explore and Explain

Wipe out!

_____ people

 Teacher Directions: Use base-ten blocks to solve the problem. Draw the blocks to show your work. There are 62 people sledding. 13 people leave to go home. 12 people go inside for hot cocoa. How many people are still sledding?

See and Show

Some word problems take two-steps to solve.

There are 87 books at the library. 23 of the books are checked out. 4 of those books are returned. How many books are at the library now?

Solve each two-step word problem.

1. Hailey has 33 books. She gives 14 books to her cousin. She gives 5 books to her sister. How many books does Hailey have left?

_____ books

I read books every day. I <u>eat</u> a few too!

2. Dale invites 25 people to his party. 8 people cannot come. 5 people come that were not invited. How many people come to the party?

_____ people

Talk Math How do you solve a two-step problem?

Name

On My Own

Solve each two-step word problem.

3. Maryanne builds a tower out of blocks.
   She uses 89 blocks. 18 blocks fall off
   the tower. Her brother knocks 14 more
   blocks off the tower. How many blocks
   are still on the tower?

_____ blocks

4. Ray has 56 comics. He gives 12 to his
   friend. He borrows 14 from his little brother.
   How many comics does Ray have now?

_____ comics

I hope Ray
lends me
that comic!

5. There are 39 penguins on the iceberg.
   11 penguins jump into the water.
   4 more penguins jump onto another
   iceberg. How many penguins are left?

_____ penguins

Problem Solving

Solve each two-step word problem.

6. There are 48 grapes in the bowl.
   Mia eats 18 grapes. Her mom adds
   22 grapes to the bowl. How many
   grapes are in the bowl now?

   _____ grapes

7. Connor's mom cut 32 apple slices for
   his two brothers and him to share. Connor
   ate 5 slices. His brothers ate a total of
   18 slices. How many apple slices were left?

   Try a kiwi slice next time!

   _____ slices

HOT Problem Write a two-step word problem using the
number sentences $13 + 10 = 23$ and $23 - 5 = 18$.

_____

_____

_____

My Homework

Homework Helper

eHelp

Need help? connectED.mcgraw-hill.com

Katie's Coffee Shop sells 84 cups of cocoa on Saturday. 23 cups are sold in the morning. 35 cups are sold in the afternoon. How many cups are sold in the evening?

$$
\begin{array}{r} 84 \\ - 23 \\ \hline 61 \end{array}
\qquad
\begin{array}{r} 61 \\ - 35 \\ \hline 26 \end{array}
$$

Helpful Hint
Some problems need to be worked in two steps.

Practice

Solve each two-step word problem.

1. 56 dogs are at the dog park. 14 dogs go home. 15 dogs come to the park. How many dogs are at the dog park now?

   _____ dogs

2. The animal shelter has 33 cats. 9 cats are adopted on Friday. 12 cats are adopted on Saturday. How many cats are left at the animal shelter?

   _____ cats

Solve each two-step word problem.

3. Mr. Henry's class has 24 students.
   3 students are absent today.
   4 students are late to school.
   How many students were on time?

   _____ students

4. Abigail has 42 erasers. She gives 18 to
   Avery. She gives 15 to Mackenzie. How
   many erasers does Abigail have left?

   _____ erasers

   Abigail
   + Mackenzie
   + Avery
   _____
   BFF

5. 15 children got on the bus. 6 children got
   off of the bus. 8 more children got on the bus.
   How many children are on the bus now?

   _____ children

Test Practice

6. Caleb had 89 race cars. He lost 23 of them. He gave 14
   to his friend. How many race cars does Caleb have left?

   75   66   52   51
   ○   ○   ○   ○

Math at Home Create a two-step word problem for your child to solve.
Have him or her show you how to solve it.

Fluency Practice

Subtract.

1.	16 − 3	2.	26 − 6	3.	13 − 8
4.	15 − 9	5.	12 − 5	6.	19 − 5
7.	11 − 9	8.	12 − 6	9.	14 − 8
10.	19 − 8	11.	18 − 9	12.	17 − 3
13.	19 − 1	14.	15 − 5	15.	17 − 4

Fluency Practice

Subtract.

1. 13
    − 6

2. 13
    − 7

3. 18
    − 8

4. 15
    − 6

5. 17
    − 5

6. 14
    − 8

7. 20
    − 9

8. 16
    − 7

9. 11
    − 10

10. 12
    − 9

11. 18
    − 7

12. 12
    − 7

13. 16
    − 2

14. 13
    − 3

15. 19
    − 7

Name ....

My Review

Vocabulary Check

Complete each sentence. Use words from the box.

subtract	fact family	regroup	difference

1. A _____ is a group of 3 numbers that work together.

2. You can _____ a ten into ten ones.

3. When you _____, you take something away from another number.

4. The answer you get from subtracting is called

   the _____.

Write an example of a fact family.

5.

___ + ___ = ___   ___ − ___ = ___

___ + ___ = ___   ___ − ___ = ___

Concept Check

Subtract.

6.
$$\begin{array}{r} 27 \\ -18 \\ \hline \end{array}$$

7.
$$\begin{array}{r} 32 \\ -17 \\ \hline \end{array}$$

8.
$$\begin{array}{r} 26 \\ -19 \\ \hline \end{array}$$

9.
$$\begin{array}{r} 48 \\ -28 \\ \hline \end{array}$$

10.
$$\begin{array}{r} 83 \\ -37 \\ \hline \end{array}$$

11.
$$\begin{array}{r} 80 \\ -26 \\ \hline \end{array}$$

12.
$$\begin{array}{r} 74 \\ -25 \\ \hline \end{array}$$

13.
$$\begin{array}{r} 71 \\ -37 \\ \hline \end{array}$$

Rewrite the numbers. Subtract.

14. $48 - 29$

15. $26 - 17$

16. $53 - 37$

Subtract. Check by adding.

17.
$$\begin{array}{r} 73 \\ -27 \\ \hline \end{array}$$
$$+ \underline{}$$

18.
$$\begin{array}{r} 63 \\ -36 \\ \hline \end{array}$$
$$+ \underline{}$$

Hit the beach!

Name _____

Problem Solving

19. Max has 34 marbles. He gives 19 marbles away. How many marbles does Max have left?

_____ marbles

20. There are 18 alligators around a pond. 10 alligators slide into the water. 2 come back out. How many alligators are around the pond now?

_____ alligators

21. Jill collects 68 golf balls. She gives 39 of them to her grandpa. She loses 8 of them. How many golf balls does Jill have left?

_____ golf balls

Test Practice

22. Matt sees 25 planes at the air show. 6 of the planes are red. 8 are white. The rest are blue. How many planes are blue?

39 ○　　　19 ○　　　14 ○　　　11 ○

Show the ways to subtract.
Check your work.

Subtract from a two-digit number.

$$25$$
$$-\ \ 4$$

Subtract with regrouping.

$$32$$
$$-\ \ 6$$

ESSENTIAL QUESTION

How can I subtract two-digit numbers?

Subtract without regrouping.

$$48$$
$$-\ 35$$

Subtract with regrouping.

$$71$$
$$-\ 19$$

Have a ball!

Let's get rolling!

Chapter 5 Place Value to 1,000

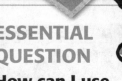

ESSENTIAL QUESTION

How can I use place value?

Look At My Collections!

Watch a video!

Watch

285

Chapter 5 Project

Number Pattern Stories

1. Each student in the group needs to complete their own story map.
2. Complete the story map and decide which number pattern your group will include in the story.
3. Use the story map to create your number pattern story.

Story Map

Title ←――――→ [ ]

Characters ←――――→ [ ]

Setting ←――――→ [ ]

Problem ←――――→ [ ]

Solution ←――――→ [ ]

Name _____

Am I Ready?

Write the number.

1.

_____

2.

_____

Compare. Write >, <, or =.

3. 27 ◯ 36

4. 15 ◯ 15

5. 45 ◯ 29

6. 12 ◯ 20

7. Tia counts the shoes in her closet. Each pair has 2 shoes. Show how she counts.

_____, 4, _____, _____, 10

How Did I Do? → Shade the boxes to show the problems you answered correctly.

| 1 | 2 | 3 | 4 | 5 | 6 | 7 |

My Math Words

Vocab abc

Review Vocabulary

ones	tens

Mark an X to show the value of the underlined digit. Then write the value of the digit.

	Tens	Ones	Value
5<u>7</u>		X	7
<u>8</u>9			
2<u>0</u>			
<u>3</u>2			
4<u>1</u>			
<u>7</u>4			
<u>6</u>8			

Write a sentence about a review word.

_____

_____

Lesson 5-7

compare

3 = 3

Lesson 5-3

digit

2 digits

Lesson 5-7

equal to (=)

400 = 400

Lesson 5-3

expanded form

536 = 500 + 30 + 6

Lesson 5-7

greater than (>)

431 > 413

Lesson 5-1

hundreds

hundreds	tens	ones

2 hundreds

Directions:
Ideas for Use

- Have students group 2 or 3 common words. Ask them to add a word that is unrelated to the group. Have them work with a friend to name the unrelated word.

- Ask students to arrange cards to show words that are opposites. Have them explain the meaning of their opposite words.

A symbol used to write numbers. The ten digits are 0, 1, 2, 3, 4, 5, 6, 7, 8, and 9.

Look at numbers to see how they are alike or different.

The representation of a number as a sum that shows the value of each digit.

Have the same value as or is the same value as.

The numbers in the range of 100–999.

The number or group with more.

Vocab Processes & Practices

Lesson 5-7

less than (<)

30   <   40

Lesson 5-1

ones

2 3

2 tens and 3 ones

20 + 3

Lesson 5-3

place value

hundreds	tens	ones
7	8	5

700 + 80 + 5 =
785

Lesson 5-1

tens

2 3

2 tens and 3 ones

20 + 3

Lesson 5-5

thousand

thousands	hundreds	tens	ones

↑
I thousand

Directions:
More Ideas for Use
- Have students arrrange the cards in alphabetical order.

- Ask students to use the blank card to write their own vocabulary card.

The numbers in the range of 0–9. It is the place value of a number.

The number or group with fewer.

The numbers in the range of 10–99. It is the place value of a number.

The value given to a digit by its place in a number.

The numbers in the range of 1,000–1,999.

My Foldable

 FOLDABLES® Follow the steps on the back to make your Foldable.

✂ -

① _____   _____   _____

② _____   _____   _____

③ _____   _____   _____

④ _____   _____   _____

Ones

Tens

Hundreds

Name

Hundreds

Explore and Explain

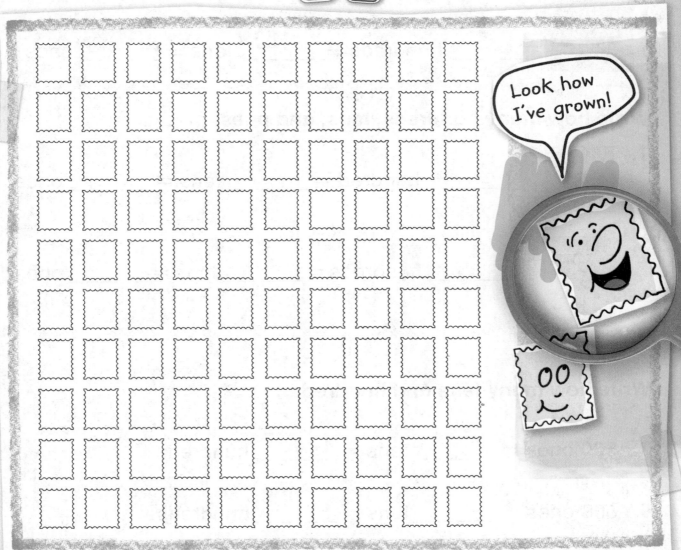

Look how I've grown!

Teacher Directions: Color each stamp yellow and count the stamps by ones.
Circle groups of ten in green to show tens. Count by tens to 100 with a classmate.
Circle one hundred stamps in red to show a hundred.

See and Show

_100_ ones   _10_ tens   I hundred

_2_ hundreds = _20_ tens = _200_ ones

Write how many hundreds, tens, and ones.

1. _____ hundred = _____ tens = _____ ones

2. _____ hundreds = _____ tens = _____ ones

Write how many tens and hundreds.

3. 500 ones = _____ tens = _____ hundreds

4. 600 ones = _____ tens = _____ hundreds

Talk Math Is 100 the same as ten tens?
How do you know?

Name

_____

On My Own

Write how many hundreds, tens, and ones.

5. _____ hundreds = _____ tens = _____ ones

6. _____ hundreds = _____ tens = _____ ones

7. _____ hundreds = _____ tens = _____ ones

Write how many tens and hundreds.

8. 800 ones = _____ tens = _____ hundreds

9. 700 ones = _____ tens = _____ hundreds

Write how many hundreds, tens, or ones.

10. 500 = _____ tens

11. 200 = _____ ones

12. 400 = _____ hundreds

13. 300 = _____ tens

14. Sara has a collection of coins. She puts them in 30 piles of 10. How many hundreds does she have?

_____ hundreds

15. Kenny places his rock collection in 20 groups of 10. How many ones does he have?

_____ ones

16. Andreas counts 300 leaves on a tree. How many tens does he count?

_____ tens

Write Math How many ones are in 4 hundreds? How do you know?

_____

_____

_____

Name _____

My Homework

Homework Helper Need help? connectED.mcgraw-hill.com

 4 hundreds = 40 tens = 400 ones

Practice

Write how many hundreds, tens, and ones.

1. _____ hundreds = _____ tens = _____ ones

2. _____ hundreds = _____ tens = _____ ones

3. _____ hundreds = _____ tens = _____ ones

Write how many tens and hundreds.

4. 900 ones = _____ tens = _____ hundreds

Write how many hundreds, tens, or ones.

5. 800 = _____ tens

6. 400 = _____ ones

7. 700 = _____ hundreds

8. 600 = _____ tens

Solve. Draw models to help if needed.

9. Katie has 5 pages of star stickers. Each page has 100 stickers. How many star stickers does Katie have in all?

_____ star stickers

10. Avante has 7 baskets of 100 beads. How many hundreds of beads does Avante have?

_____ hundreds

Vocabulary Check

11. Circle the digit that shows **hundreds**. Underline the **ones** digit once. Underline the **tens** digit twice.

600

Math at Home Ask your child to count by hundreds to 1,000.

300   Chapter 5 • Lesson 1

Hundreds, Tens, and Ones

Lesson 2

ESSENTIAL QUESTION
How can I use place value?

 I love collecting cans and jumping over them too!

Explore and Explain

Watch ▶

Tools

hundreds	tens	ones

_____ cans

 Teacher Directions: Mya collected cans to recycle. The amount she collected had 3 hundreds, 6 tens, and 7 ones. Use base-ten blocks to show the number in the chart. Write the number.

See and Show

There are 427 pennies in this jar. Use hundreds, tens, and ones to show 427.

Hey, I'm full!

hundreds	tens	ones
4	2	7

427 four hundred twenty-seven

Write how many hundreds, tens, and ones. Then write the number.

1. 2 , 6 , and 3

hundreds	tens	ones

_____

2. 5 , 1 , and 8

hundreds	tens	ones

_____

Talk Math What is the value of the 1 in 712, in 165, and in 381?

Name _____

On My Own

Write how many hundreds, tens, and ones.
Then write the number.

3. 3 [image], 8 [image], and 2 [image]

hundreds	tens	ones

_____

4. 6 [image], 4 [image], and 3 [image]

hundreds	tens	ones

_____

5. 7 [image], 0 [image], and 9 [image]

hundreds	tens	ones

_____

6. 7 [image], 2 [image], and I [image]

hundreds	tens	ones

_____

Write how many hundreds, tens, and ones.

7. 684 = _____ hundreds, _____ tens, and _____ ones

8. 805 = _____ hundreds, _____ tens, and _____ ones

9. 290 = _____ hundreds, _____ tens, and _____ ones

Problem Solving

10. This number has 4 hundreds and 2 ones. The number of tens is the digit between the number of ones and the number of hundreds. What is the number?

_____

You can view the largest, most complete T. Rex at The Field Museum in Chicago, Illinois. It has more than 200 bones. Look at the number 200.

11. What number is in the hundreds place?

_____

12. What number is in the tens place?

_____

13. What number is in the ones place?

_____

Grrrrr!

   What do hundreds, tens, and ones tell you about a number?

_____

_____

_____

Name _____

My Homework

Homework Helper Need help? connectED.mcgraw-hill.com

Use hundreds, tens, and ones to show a number.

2 _____, 3 _____, and 5 _____

hundreds	tens	ones
2	3	5

Practice

Write how many hundreds, tens, and ones.
Then write the number.

1. 3 _____, 2 _____, and 4 _____

hundreds	tens	ones

_____

2. 1 _____, 3 _____, and 8 _____

hundreds	tens	ones

_____

3. 4 _____, 1 _____, and 3 _____

hundreds	tens	ones

_____

Write how many hundreds, tens, and ones.

4. 536 = _____ hundreds, _____ tens, and _____ ones

5. 295 = _____ hundreds, _____ tens, and _____ ones

Solve. Draw base-ten blocks to model.

6. Percy has 342 pencils.
   How many tens does he have?

_____ tens

7. Luis collected 613 beads.
   How many hundreds does he have?

_____ hundreds

Test Practice

8. A number has 6 tens, 9 ones, and 5 hundreds.
   What number is it?

   695   596   569   965
   ○   ○   ○   ○

Math at Home Write the number 647. Ask your child to tell you how many hundreds, tens, and ones are in the number.

Name _____

Place Value to 1,000

Lesson 3

ESSENTIAL QUESTION
How can I use place value?

Explore and Explain

I love my **button** collection!

hundreds	tens	ones

_____ + _____ + _____ = _____

 Teacher Directions: Roll a red number cube. Write that number in the hundreds place. Roll again. Write that number in the tens place. Roll again. Write that number in the ones place. Write the value of each number and their sum when added together.

Copyright © The McGraw-Hill Companies, Inc. (tr)The McGraw-Hill Companies

See and Show

Place value tells the value of a **digit** in a number. To write a number in **expanded form**, show the value of each digit.

hundreds	tens	ones
2	1	3

Helpful Hint

When you write a number in expanded form, you show the value of each digit.

2 hundreds 1 ten   3 ones

$200 + 10 + 3 = 213$

Write each number in expanded form.
Then write the number.

1. 1 hundred, 4 tens, and 9 ones

_____ + _____ + _____ = _____

2. 3 hundreds, 2 tens, and 5 ones

_____ + _____ + _____ = _____

3. 5 hundreds, 7 tens, and 1 one

_____ + _____ + _____ = _____

 How are 562 and 265 the same?
How are they different?

Name _____

On My Own

**Write each number in expanded form.
Then write the number.**

4. 7 hundreds, 0 tens, and 2 ones

_____ + _____ + _____ = _____

5. 4 hundreds, 2 tens, and 1 one

_____ + _____ + _____ = _____

Circle the value of the green digit.

6. 965

900   90   9

7. 673

300   30   3

Write the number.

8. 800 + 30 + 3 = _____

9. 200 + 90 + 0 = _____

10. 700 + 10 + 9 = _____

11. 900 + 80 + 9 = _____

Write the number in expanded form.

12. 254 = _____ + _____ + _____

13. 526 = _____ + _____ + _____

14. The number of raisins in a box has 4 in the ones place, 2 in the hundreds place, and 5 in the tens place. How many raisins are in the box?

_____ raisins

15. Sharon has a pile of 100 erasers, a pile of 20 erasers, and a pile of 5 erasers. How many erasers does Sharon have in all?

Erasers make me smile!

_____ erasers

16. Luis wrote a story. The number of pages in his story has 1 hundred, 7 tens, and 9 ones. How many pages are in his story?

_____ pages

HOT Problem A number has three digits. Which digit has the greatest value and why?

_____

_____

_____

Name _____

My Homework

Homework Helper

Need help? connectED.mcgraw-hill.com

You can write a number in expanded form by writing the value of each digit.

hundreds	tens	ones
2	6	3

2 hundreds	6 tens	3 ones
200	+ 60	+ 3 = 263

Practice

**Write each number in expanded form.
Then write the number.**

1. 2 hundreds, 6 tens, and 8 ones

   _____ + _____ + _____ = _____

2. 3 hundreds, 1 ten, and 5 ones

   _____ + _____ + _____ = _____

3. 6 hundreds, 0 tens, and 2 ones

   _____ + _____ + _____ = _____

Write the number in expanded form.

4. $637 = $ _____ + _____ + _____

5. $742 = $ _____ + _____ + _____

6. $295 = $ _____ + _____ + _____

Here's what I'll look like on my baseball card!

14

Joe

7. Joe has 310 baseball cards. He writes the number of cards in expanded form. How many ones will he write?

_____ ones

8. Trey reads that 482 people went to the baseball game. How can Trey show this number of people in expanded form?

_____ + _____ + _____

Vocabulary Check

Circle the numbers that show the word below.

9. expanded form $300 + 42$ $300 + 40 + 2$ $3 + 4 + 2$

 Math at Home Have your child tell you a three-digit number. Then ask your child to tell you the value of the first digit.

Name

Vocabulary Check

Draw lines to match.

1. **hundreds** $500 + 60 + 3$

2. **expanded form**

hundreds	tens	ones
5	6	3

3. **place value**   3

4. **digit**   563
                                          ↑

Concept Check

Write how many tens and hundreds.

5. 900 ones = _____ tens = _____ hundreds

6. 600 ones = _____ tens = _____ hundreds

Write how many hundreds, tens, or ones.

7. 300 = _____ ones   8. 400 = _____ tens

Write how many hundreds, tens, and ones.
Then write the number.

9. 8 ___ , 2 ___ , 1 ___

hundreds	tens	ones

_____

Write each number in expanded form.
Then write the number.

10. five hundred sixty-six

_____ + _____ + _____ = _____

11. nine hundred ten

_____ + _____ + _____ = _____

12. Janie is thinking of a number. It has 8
ones and 9 tens. It has the same amount
of hundreds as tens. What is the number?

_____

Test Practice

13. A number has a 5 in the tens place. The number
of ones is 3 less than the number of tens. The
number of hundreds is 1 less than the number
of ones. What is the number?

512 ○   215 ○   152 ○   125 ○

Name ....

Problem Solving
STRATEGY: Use Logical Reasoning

Watch ▶

Eva's house number has three digits.
The sum of the digits is 7. Two of the
digits are even. The middle digit is 1.
The first digit is less than the last digit.
What is Eva's house number?

1 Understand
Underline what you know.
Circle what you need to find.

2 Plan
How will I solve the problem?

3 Solve
Use logical reasoning.

First digit plus last digit equals ___6___ .

___2___ and ___4___ are even and equals ___6___ .

___2___ is less than ___4___ .

Eva's house number is ___214___ .

4 Check
Is my answer reasonable? Explain.

Practice the Strategy

The number of dogs at the shelter is a two-digit, odd number. It is more than 80. The tens digit is less than the ones digit. The sum of the digits is 17.

How many dogs are at the shelter?

1 Understand Underline what you know. Circle what you need to find.

2 Plan How will I solve the problem?

3 Solve I will...

_____ dogs

4 Check Is my answer reasonable? Explain.

Good night!

Apply the Strategy

1. A number has three digits. The digit in the hundreds place is the difference of 4 and 2. The digit in the tens place is the sum of 2 and 3. The digit in the ones place is 2 more than the digit in the tens place. What is the number?

        _____

2. Sylvia writes this number: three ones, two tens, and seven hundreds. What number is it?

        _____

3. Ruby is trying to solve a number riddle. The digit in the tens place is less than 2. It is an odd number. The number in the hundreds place is greater than 8. The number in the ones place is the sum of 6 and 3. What is the number?

I LOVE riddles!

        _____

Review the Strategies

Choose a strategy
- Make a model.
- Write a number sentence.
- Find a pattern.

4. Paco puts pennies into three groups. He counts 80 pennies in the first group, 3 pennies in the second group, and 500 pennies in the third group. How many pennies does he have?

_____ pennies

5. Ella's family is going to visit her cousins. They will drive for 3 days. Each day they will drive 60 miles. How many miles will they drive in all?

_____ miles

6. Jamie ate 20 cherries on Monday. She ate 20 cherries on Tuesday. She ate 20 cherries on Wednesday. If she continues, how many cherries will Jamie have eaten by Thursday?

_____ cherries

My Homework

Pat's Fruit Stand has 534 peaches for sale. Pat sells 30 peaches this afternoon. How many peaches are left to sell?

eHelp

Peachy!

1 Understand Underline what you know.
Circle what you need to find.

2 Plan How will I solve the problem?
I will use logical reasoning.

3 Solve $534 = 500 + 30 + 4$

30 peaches were sold.

$504 = 500 + \cancel{30} + 4$

There are 504 peaches left.

4 Check Is my answer reasonable?

Underline what you know. Circle what you need to find.

1. Lucy is thinking of a number. The number is greater than two hundred twenty-five. Her number is less than 2 hundreds, 2 tens, and 7 ones. What is Lucy's number?

   _____

2. The number of blocks has 9 in the ones place. The number in the hundreds place is one more than the number in the tens place. Those two numbers equal 11. How many blocks are there?

   _____ blocks

3. Ralph has 957 race car stickers. He gave 10 to Ken. He gave 100 to Matt. How many stickers does Ralph have left?

   _____ stickers

4. My class collected 389 milk jugs. We used 40 jugs. We gave 30 milk jugs to the first graders. How many milk jugs did we have left?

Math at Home Ask your child to write the number 961 on a piece of paper. Tell him or her to write the number that is one less than that number on the paper in expanded form.

   _____ milk jugs

Name _____

Read and Write Numbers to 1,000

Explore and Explain ▶ Watch

101	102	103	104	105	106	107	108	109	110
111	112	113	114		116	117	118	119	120
121	122	123	124	125	126	127	128	129	130
131		133	134	135	136	137	138	139	140
141	142	143	144	145	146	147	148	149	150
151	152	153	154	155	156		158	159	160
161	162	163	164	165	166	167	168	169	170
171	172	173	174	175	176	177		179	180
181	182	183	184	185	186	187	188	189	190
191	192	193	194	195	196	197	198		200

 Teacher Directions: Fill in the missing numbers. Read the numbers from 101 to 150. Color those numbers yellow. Read the numbers from 151 to 200. Color those numbers green.

See and Show

Helpful Hint
One thousand is equal to ten hundreds. We write this as 1,000.

You can read and write numbers with symbols and words.

1 one	11 eleven	10 ten	100 one hundred
2 two	12 twelve	20 twenty	200 two hundred
3 three	13 thirteen	30 thirty	300 three hundred
4 four	14 fourteen	40 forty	400 four hundred
5 five	15 fifteen	50 fifty	500 five hundred
6 six	16 sixteen	60 sixty	600 six hundred
7 seven	17 seventeen	70 seventy	700 seven hundred
8 eight	18 eighteen	80 eighty	800 eight hundred
9 nine	19 nineteen	90 ninety	900 nine hundred
			1,000 one **thousand**

Read five hundred thirty-eight.

Write the number. ___538___

Read the number. Write the number.

1. thirty-eight

2. one hundred twenty-one

_____

_____

Write the number in words. Read the number.

3. 710 _____

4. 900 _____

 Explain how you would write 62 and 602 using words.

Name ....

On My Own

Read the number.
Write the number.

5. one hundred ninety-nine

_____

6. seven hundred twenty-eight

_____

7. three hundred ten

_____

8. two hundred eighty-five

_____

9. nine hundred seventy-seven

_____

10. four hundred sixty-four

_____

Write the number in words. Read the number.

11. 1,000 _____

12. 718 _____

13. 614 _____

14. 244 _____

15. 321 _____

Problem Solving

16. There are 119 frogs in the pond. Write the number of frogs in the pond in words.

_____

17. Amber is thinking of this number word: two hundred fifty-six. What is the number?

_____

18. There are four hundred twenty-seven geese in the lake. Write the number of geese in the lake.

Which way to the lake?

_____ geese

Write Math   Why is it important to know numbers and number names?

_____

_____

_____

Name _____

My Homework

Homework Helper **Need help?** connectED.mcgraw-hill.com

Nine hundred fifty-four in numbers = 954

467 = four hundred sixty-seven

Practice

Read the number. Write the number.

1. two hundred thirty-five

   _____

2. six hundred seventy-two

   _____

3. one hundred forty-six

   _____

4. three hundred twenty-one

   _____

Write the number in words.
Read the number.

5. 682 _____

6. 431 _____

Solve. Draw base-ten blocks if needed.

7. There are 429 students at Linden School. Cora wants to write the number in words to put in a newsletter. What should she write?

_____ students

8. Marco lives at nine hundred thirty-one Maple Street. Write the number.

_____

9. Rani knows that there are three hundred sixty-five days in one year. How can she write this number?

_____ days

Vocabulary Check

10. Circle the number that shows **thousands**.

    1,365   1,000   3,651
     ↑   ↑   ↑

Math at Home Say three hundred forty-seven and have your child write the number.

Count by 5s, 10s, and 100s

Lesson 6

ESSENTIAL QUESTION
How can I use place value?

Explore and Explain

201	202	203	204	205	206	207	208	209	210
211	212	213	214	215	216	217	218	219	220
221	222	223	224	225	226	227	228	229	230
231	232	233	234	235	236	237	238	239	240
241	242	243	244	245	246	247	248	249	250
251	252	253	254	255	256	257	258	259	260
261	262	263	264	265	266	267	268	269	270
271	272	273	274	275	276	277	278	279	280
281	282	283	284	285	286	287	288	289	290
291	292	293	294	295	296	297	298	299	300

 Teacher Directions: Skip count by 5. Color those numbers yellow. Skip count by 10. Circle those numbers red. If the chart continued, and you counted by hundreds, what would be the next number you would say?

See and Show

Number patterns can help you count. In these patterns, each number is more.

Helpful Hint
Patterns can also be 5 less, 10 less, or 100 less.

560
555
550
545

Count by **fives**.
Each number
is __5__ more.

575
565
555
545

Count by **tens**.
Each number
is __10__ more.

845
745
645
545

Count by **hundreds**.
Each number
is __100__ more.

ENGINE HOUSE #1

Write the missing numbers. Then write the counting pattern.

1. 340, 350, 360, ____, 380

The pattern is _____ .

2. 575, 580, ____, 590, ____

The pattern is _____ .

3. 941, 841, ____, 641, 541

The pattern is _____ .

4. 680, 675, ____, ____, 660

The pattern is _____ .

Talk Math How can you tell if a number pattern is counting by hundreds?

Name _____

On My Own

Write the missing numbers.
Then write the counting pattern.

I will make the missing numbers appear!

5. 500, 510, ____, 530, ____

The pattern is _____.

6. 310, ____, 320, ____, 330

The pattern is _____.

7. 800, 790, 780, ____, 760

The pattern is _____.

8. 655, ____, 455, 355, ____

The pattern is _____.

9. ____, 486, 586, 686, ____

The pattern is _____.

10. 500, 600, 700, ____, 900

The pattern is _____.

11. 234, ____, ____, 264, 274

The pattern is _____.

12. ____, 515, 510, 505, ____

The pattern is _____.

Problem Solving

Processes
&Practices

13. Paul counts by hundreds. He starts with the number 123. Write the numbers Paul counts.

123, _____, _____, _____, _____, _____

14. Ali counts by tens. She starts with 325. Ali counts 325, 335, 345, 365, 385. What numbers did Alli leave out?

What's next?

_____ and _____

15. Write the numbers you would say if you count backwards by fives. Start with 230 and end at 200.

230, _____, _____, _____, _____, _____, 200

Write Math Felicia counts from 300 to 400. Would it be faster to count by 5s or by 10s? Explain.

_____

_____

_____

My Homework

Homework Helper **Need help?** connectED.mcgraw-hill.com

Number patterns can help you count.

Count by fives: 40, 45, 50, 55, 60, 65

Count by tens: 120, 130, 140, 150, 160

Count by hundreds: 200, 300, 400, 500, 600

Practice

Write the missing numbers. Then write the counting pattern.

1. 1,000, 995, ____, 985, ____

   The pattern is _____.

2. 524, ____, 544, 554, ____

   The pattern is _____.

3. ____, 283, 383, ____, 583

   The pattern is _____.

4. ____, 843, 743, 643, ____

   The pattern is _____.

5. 420, 430, ____, 450, ____

   The pattern is _____.

6. 525, ____, 535, ____, 545

   The pattern is _____.

Continue the counting pattern.

7. 835, 840, 845, _____, _____, _____

8. 410, 420, 430, _____, _____, _____

9. 900, 800, 700, _____, _____, _____

10. Shari counts: 169, 159, 149, 139, 129. Shari wants Miguel to guess her counting pattern. What should Miguel guess?

The pattern is _____.

Test Practice

11. Mark the number pattern that shows 10 less.

820, 830, 840   980, 970, 960
   ○   ○

923, 922, 921   400, 500, 600
   ○   ○

Math at Home Pick a three-digit number. Ask your child to count by ones, tens, and/or hundreds.

Compare Numbers to 1,000

Lesson 7

ESSENTIAL QUESTION
How can I use place value?

Explore and Explain

How much are you worth?

A lot, I'm sure!

>

 Teacher Directions: Use base-ten blocks to show 331 and 208 on the mat. Place them on the appropriate side of the symbol. Draw the base-ten blocks.

See and Show

Use these steps to compare numbers.

Step 1 Line up the numbers by place value in the chart.

Step 2 **Compare** the digits in the greatest place value first. If they are the same, compare digits in the next place value.

greater than		
hundreds	tens	ones
6	1	8
6	3	4

634 > 618

634 is *greater than* 618.

less than		
hundreds	tens	ones
8	4	6
8	2	3

823 < 846

823 is *less than* 846.

equal to		
hundreds	tens	ones
9	7	8
9	7	8

978 = 978

978 is *equal to* **978**.

Compare. Write >, <, or =.

1. 142 ◯ 124

2. 253 ◯ 257

3. 313 ◯ 313

4. 842 ◯ 795

5. 694 ◯ 694

6. 203 ◯ 153

7. 100 ◯ 1,000

8. 999 ◯ 99

9. 133 ◯ 133

10. 743 ◯ 734

11. 861 ◯ 871

12. 542 ◯ 452

Talk Math Explain how to compare 567 and 576.

Name _____

On My Own

Compare. Write >, <, or =.

Helpful Hint
Compare the digits in the greatest place value first.

13. 150 ◯ 150

14. 132 ◯ 213

15. 689 ◯ 627

16. 425 ◯ 425

17. 907 ◯ 899

18. 533 ◯ 533

19. 207 ◯ 210

20. 697 ◯ 667

21. 411 ◯ 421

22. 619 ◯ 621

23. 729 ◯ 729

24. 325 ◯ 300

25. 332 ◯ 335

26. 984 ◯ 894

Circle the number that is less than the purple number.

27.   568

        588   464

28.   409

        410   406

Circle the number that is greater than the purple number.

29.   311

        313   211

30.   653

        651   655

31. I am greater than 3 hundreds, 2 tens, and 2 ones. I am less than 3 hundreds, 2 tens, 4 ones. What number am I?

Hey, someone bit me!

_____

32. Ravi is making cookies. He wants to have two equal groups of cookies. He made 45 cookies for one group. How many cookies does Ravi need for the other group?

_____ cookies

33. Demi collected 124 seashells. Eva collected 142 seashells. Who collected less seashells?

_____

Write Math   Explain how you decide which number is greater than the other.

_____

_____

_____

Name _____

My Homework

Homework Helper   Need help? connectED.mcgraw-hill.com

greater than		
hundreds	tens	ones
5	4	6
4	6	2

546 > 462

less than		
hundreds	tens	ones
3	2	5
5	1	8

325 < 518

equal to		
hundreds	tens	ones
2	5	8
2	5	8

258 = 258

Practice

Compare. Write >, <, or =.

1. 415 ◯ 451

2. 623 ◯ 678

3. 730 ◯ 830

4. 375 ◯ 375

5. 549 ◯ 560

6. 258 ◯ 239

7. 109 ◯ 111

8. 382 ◯ 379

9. 445 ◯ 545

10. 272 ◯ 275

11. 818 ◯ 816

12. 357 ◯ 357

Solve to answer each question below.

13. Uri has 529 bugs in his collection.
Elena has 513 bugs in her collection.
Who has a greater number of bugs?

_____

14. Seth saves 347 bottle caps.
Jorge saves 345. Who saves
the greater number of bottle caps?

_____

15. Elisa has 125 paper dolls in her collection.
Katherine has 138 paper dolls in her
collection. Who has less dolls?

_____

Vocabulary Check

Draw lines to match.

16. **greater than**	=
17. **less than**	>
18. **equal to**	<

Math at Home Ask your child to name numbers that are greater than,
less than, and equal to 807.

Name _____

My Review

Vocabulary Check

Draw lines to match.

1.	**expanded form**	498 ↑
2.	**hundreds**	>
3.	**equal to**	1,000 ↑
4.	**greater than**	=
5.	**less than**	700 + 30 + 6
6.	**thousand**	<

Concept Check

Write how many hundreds, tens, and ones.

7.

_____ hundreds = _____ tens = _____ ones

Write how many hundreds, tens, and ones.
Then write the number.

8.

2 ▨ , 3 ▮ , and 9 ▪

hundreds	tens	ones

_____

Write the number in expanded form.
Then write the number.

9. 5 hundreds, 6 tens, and 0 ones

_____ + _____ + _____ = _____

10. 8 hundreds, 7 tens, and 8 ones

_____ + _____ + _____ = _____

Write the number.

11. six hundred twenty-nine   12. five hundred one

_____   _____

Compare. Write >, <, or =.

13. 100 ◯ 1,000   14. 468 ◯ 406

15. 711 ◯ 711   16. 394 ◯ 294

Problem Solving

17. Malika has 9 sheets of stickers. Tonya has 7 sheets of stickers. 100 stickers are on each sheet. Who has more stickers?

_____

18. Show another way to write six hundred twenty-three.

_____ + _____ + _____

19. Sam has 998 markers. Write the number words to show how many markers.

_____ markers

Test Practice

20. Leah has 368 sunflower seeds. How many hundreds does she have?

   3   6   8   368

   ○   ○   ○   ○

Show the ways you can use place value.

Write the number.

seven hundred sixty-three

_____

Write the number in expanded form.

233

_____ + _____ + _____

ESSENTIAL QUESTION

How can I use place value?

Compare.

763 ◯ 637

Write the missing numbers.

560, _____, 580, _____

Ready?
Set?
Solve!

6 Add Three-Digit Numbers

ESSENTIAL QUESTION
How can I add three-digit numbers?

MAIN ST.

LITTLE RD.

This is My Town

Watch a video!

Watch

Chapter 6 Project

Practice Game

1. Work together to come up with a math game that involves adding three-digit numbers using game cards or a board game.

2. Decide what to name the game, all of the materials you will need to play the game, and explain directions on how to play the game below.

Name of Game: _____

Materials: _____

Game Instructions: _____

_____

_____

_____

_____

_____

_____

_____

_____

Name _____

Am I Ready?

Add.

1.   9
      + 8
      ————

2.   8
      + 5
      ————

3.   40
      +30
      ————

4.   56
      +28
      ————

5.   8
      + 8
      ————

6.   14
      + 5
      ————

7.   60
      +30
      ————

8.   83
      +12
      ————

9.   27
      +18
      ————

10. Destiny eats 21 peanuts on Friday.
    She eats 29 peanuts on Saturday.
    How many peanuts did she eat in all?

    _____ peanuts

How Did I Do?

Shade the boxes to show the problems you answered correctly.

1	2	3	4	5	6	7	8	9	10

My Math Words

Review Vocabulary

ones	regroup	sum	tens

Use the review words to tell what the example shows about the word problem.

```
   2 9
+ 1 6
_____
```

```
   2 9
+ 1 6
_____
```

There are 29 houses on my street.
There are 16 houses on my friend's
street. How many houses in all?

```
  1
  29
+16
_____
```

```
  29
+16
_____
  45
_____
```

My Vocabulary Cards

Directions:
Ideas for Use

- Have students use the blank cards to write words from previous chapters that they would like to review.

- Ask students to use the blank cards to write their own vocabulary cards.

My Foldable

FOLDABLES Follow the steps on the back to make your Foldable.

✂

$$\begin{array}{r} 145 \\ + 112 \\ \hline \end{array}$$

$$\begin{array}{r} 145 \\ + 109 \\ \hline \end{array}$$

$$\begin{array}{r} 145 \\ + 155 \\ \hline \end{array}$$

145

145

145

145
+ 155

109
+ 601

112
+ ZII

Name

Explore and Explain

Gee, how do I get to G Street?

$97 + 16 =$ _____

_____ + _____ = _____

Teacher Directions: Model using base-ten blocks. My town has 97 streets named by numbers and 16 streets named by letters of the alphabet. How many streets are named by numbers and letters of the alphabet? Take the ones from 16 and add them to 97 to make 100. Write the new problem. Add.

See and Show

You can make a hundred to help you add.

Find 199 + 43.

Take apart 43 to make 1 + 42.
Then add 1 + 199. Add 42 to that sum.

199 + 43

199 + 1 + 42

200 + 42 = 242

So, 199 + 43 = 242.

Helpful Hint
1 + 199 = 200

Make a hundred to add.

1.   77 + 298

_____ + _____ + 298

75 + _____ = _____

So, 77 + 298 = _____ .

2.   237 + 99

_____ + _____ + 99

236 + _____ = _____

So, 237 + 99 = _____ .

Talk Math   Explain how you know which addend to take apart to make a hundred to add.

Name _____

On My Own

Make a hundred to add.

3. 98 + 393

98 + _____ + _____

_____ + 391 = _____

So, 98 + 393 = _____.

4. 45 + 199

_____ + _____ + 199

44 + _____ = _____

So, 45 + 199 = _____.

5. 99 + 468

99 + _____ + _____

_____ + 467 = _____

So, 99 + 468 = _____.

6. 62 + 199

_____ + _____ + 199

61 + _____ = _____

So, 62 + 199 = _____.

7. 99 + 798

99 + _____ + _____

_____ + 797 = _____

So, 99 + 798 = _____.

8. 94 + 198

_____ + _____ + 198

92 + _____ = _____

So, 94 + 198 = _____.

9. There are 299 animals at the fair. 76 more animals are brought to the fair. How many animals are at the fair now?

I'm the prettiest pig at the fair!

_____ animals

10. There are 334 people at the town fair. 89 more people come to the fair. How many people are at the fair in all?

_____ people

11. Leon is adding a number to 35. He takes apart 35 into 33 and 2 so he can use the 2 to make a hundred. Is Leon adding 199 or 198 to 35?

_____

Write Math   How can taking apart a number help you add to a three-digit number?

_____

_____

_____

My Homework

Homework Helper Need help? connectED.mcgraw-hill.com

You can make a hundred to add.

Find $197 + 29$.

Take apart 29 to make $26 + 3$.
Then add 197 and 3 to make 200.

Helpful Hint
Take apart a number to make a hundred. Then add.

$197 + 29$

$197 + 3 + 26$

$200 + 26 = 226$ So, $197 + 29 = 226$.

Practice

Make a hundred to add.

1. $23 + 398$

_____ + _____ + 398

$21 +$ _____ $=$ _____

So, $23 + 398 =$ _____.

2. $178 + 98$

_____ + _____ + 98

$176 +$ _____ $=$ _____

So, $178 + 98 =$ _____.

Make a hundred to add.

3. 77 + 196

_____ + _____ + 196

73 + _____ = _____

So, 77 + 196 = _____.

4. 245 + 99

_____ + _____ + 99

244 + _____ = _____

So, 245 + 99 = _____.

5. 197 people ride the bus in the morning. 74 people ride the bus in the evening. How many people ride the bus in all?

Next stop, Lesson 2!

_____ people

Test Practice

6. How could you take apart 128 to solve this addition number sentence?

$$128 + 94 = \underline{\hspace{1cm}}$$

108 + 20 ○

100 + 28 ○

122 + 6 ○

120 + 8 ○

Math at Home Ask your child to solve 99 + 598 by taking apart an addend.

Name _____

Add Hundreds

Explore and Explain

WOW!

Say ah!

_____ + _____ = _____ patients

 Teacher Directions: Model using base-ten blocks. The doctor in my town saw 200 patients last month. This month he has seen 100 patients. How many patients has the doctor seen in all? Write the addition sentence.

See and Show

You can use addition facts to add hundreds.

Find 300 + 400.

> **Helpful Hint**
> You know that
> 3 + 4 = 7, so
> 300 + 400 = 700.

$$\begin{array}{r} 3 \text{ hundreds} \\ + \ 4 \text{ hundreds} \\ \hline 7 \text{ hundreds} \end{array}$$

$300 + 400 = \underline{700}$

Add.

1. $600 + 100 = $ _____

2. $300 + 300 = $ _____

3. $200 + 300 = $ _____

4. $700 + 200 = $ _____

5. $400 + 200 = $ _____

6. $800 + 100 = $ _____

7. $\begin{array}{r} 200 \\ + \ 100 \\ \hline \end{array}$

8. $\begin{array}{r} 100 \\ + \ 300 \\ \hline \end{array}$

9. $\begin{array}{r} 500 \\ + \ 200 \\ \hline \end{array}$

 What addition fact can help you find 600 + 100? Explain.

On My Own

Add.

10. 200 + 300 = _____

11. 400 + 400 = _____

12. 700 + 100 = _____

13. 600 + 300 = _____

14. 400 + 200 = _____

15. 100 + 100 = _____

16.
```
   300
+ 400
```

17.
```
   100
+ 800
```

18.
```
   200
+ 600
```

19.
```
   500
+ 300
```

20.
```
   200
+ 700
```

21.
```
   400
+ 300
```

22.
```
   600
+ 200
```

23.
```
   300
+ 100
```

24.
```
   900
+   0
```

25.
```
   700
+ 100
```

Here's the news... You can add 100s!

26. Ava counts 100 cars in a parking lot. She counts 200 cars in another parking lot. How many cars did Ava count in all?

_____ cars

27. A baker bakes 400 bagels on Saturday. He bakes 200 bagels on Sunday. How many bagels does he bake on those two days?

_____ bagels

28. Jan delivers 300 newspapers on Friday. She delivers 400 newspapers on Sunday. How many newspapers did she deliver in all?

_____ newspapers

HOT Problem Pam has collected 100 pull-tabs for charity. Jake has 100 of them and Pat has 200. How many do they have altogether? Tell how knowing addition facts can help you solve this problem. Solve.

_____

_____

_____

Name _____

My Homework

Homework Helper Need help? connectED.mcgraw-hill.com

You can use addition facts to add hundreds.

Find 200 + 500.

> **Helpful Hint**
> Think of addition facts you know. They can help you add hundreds.

$$\begin{array}{r} 2 \text{ hundreds} \\ + 5 \text{ hundreds} \\ \hline 7 \text{ hundreds} \end{array}$$

200 + 500 = 700

Practice

Add.

1. 200 + 300 = _____

2. 500 + 300 = _____

3. 100 + 100 = _____

4. 600 + 100 = _____

5. $\begin{array}{r} 700 \\ + 200 \\ \hline \end{array}$

6. $\begin{array}{r} 800 \\ + 100 \\ \hline \end{array}$

7. $\begin{array}{r} 400 \\ + 400 \\ \hline \end{array}$

Add.

8.   100
    + 200
	9. 500
	    + 100
		10. 200
		    + 400

11.   100
    + 700
	12. 100
	    + 400
		13. 300
		    + 300

14.   700
    + 200
	15. 500
	    + 400
		16. 200
		    + 600

17. Zoe counts 300 seeds in her pumpkin.
Michael counts 500 seeds in his pumpkin.
How many seeds do they count in all?

_____ seeds

Test Practice

18. Find the sum.

$$600 + 300 = \underline{\hspace{1cm}}$$

900   300   9   3
○   ○   ○   ○

Math at Home Ask your child how knowing 4 + 5 = 9 helps him or her find 400 + 500.

Mentally Add 10 or 100

Explore and Explain

The kids will love this field trip!

_____ students

_____ students

Teacher Directions: Model using base-ten blocks. 126 students are waiting in line for the yellow buses and 126 students are also waiting for the green buses. 100 more students get in line for the yellow buses. 10 more students get in line for the green buses. How many students are in line for each bus now? Write the numbers.

See and Show

To add 10 or 100, think of addition facts you know.

Mentally add 100.

$$215 + 100$$

Helpful Hint
You know that
$2 + 1 = 3$.

$$215 + 100 = 315$$

Mentally add 10.

$$250 + 10$$

Helpful Hint
You know that
$5 + 1 = 6$.

$$250 + 10 = 260$$

Add.

1. $$450 + 100$$

2. $$630 + 10$$

3. $$801 + 100$$

4. $$115 + 10$$

5. $$500 + 10$$

6. $$237 + 100$$

7. $327 + 100 =$ _____

8. $755 + 10 =$ _____

Talk Math   Why is it easy to mentally add 10 or 100?

Name

On My Own

Add.

9.
$$122 + 100$$

10.
$$378 + 10$$

11.
$$500 + 10$$

12.
$$200 + 100$$

13.
$$747 + 100$$

14.
$$800 + 10$$

15.
$$400 + 10$$

16.
$$668 + 100$$

17.
$$817 + 100$$

Find the missing addend.

18. $392 + \rule{2cm}{0.4pt} = 492$

19. $\rule{2cm}{0.4pt} + 10 = 825$

20. $101 + \rule{2cm}{0.4pt} = 111$

21. $756 + \rule{2cm}{0.4pt} = 856$

22. $\rule{2cm}{0.4pt} + 100 = 823$

23. $\rule{2cm}{0.4pt} + 10 = 846$

24. $385 + \rule{2cm}{0.4pt} = 485$

Hi neighbor!

Problem Solving

25. 448 birds are in the park. 100 more birds fly into the park. How many birds are in the park now?

_____ birds

26. A mail carrier delivered 352 packages last month. She delivered 10 more packages last week. How many packages did she deliver in all?

Do you like my hair style?

_____ packages

27. There are 112 sheep in a barn. More sheep come into the barn. Now there are 122 sheep. How many sheep came into the barn?

_____ sheep

Write Math   Explain how to add 100 to 899.

_____

_____

_____

Name

My Homework

Homework Helper Need help? connectED.mcgraw-hill.com

Mentally add 100.

$$
\begin{array}{r}
152 \\
+\ 100 \\
\hline
252
\end{array}
$$

Helpful Hint
You know that
1 + 1 = 2.

Mentally add 10.

$$
\begin{array}{r}
152 \\
+\ 10 \\
\hline
162
\end{array}
$$

Helpful Hint
You know that
5 + 1 = 6.

Practice

Add.

1.
$$
\begin{array}{r}
235 \\
+\ 100 \\
\hline
\end{array}
$$

2.
$$
\begin{array}{r}
370 \\
+\ 10 \\
\hline
\end{array}
$$

3.
$$
\begin{array}{r}
532 \\
+\ 100 \\
\hline
\end{array}
$$

4.
$$
\begin{array}{r}
365 \\
+\ 10 \\
\hline
\end{array}
$$

5.
$$
\begin{array}{r}
178 \\
+\ 100 \\
\hline
\end{array}
$$

6.
$$
\begin{array}{r}
600 \\
+\ 10 \\
\hline
\end{array}
$$

Find the missing addend.

7. $389 + \underline{\hspace{1.5cm}} = 489$

8. $\underline{\hspace{1.5cm}} + 10 = 235$

9. $634 + \underline{\hspace{1cm}} = 644$

10. $295 + \underline{\hspace{1.5cm}} = 395$

11. Jackson has 389 ants in his ant farm. He puts 10 more ants in the farm. How many ants are in his ant farm now?

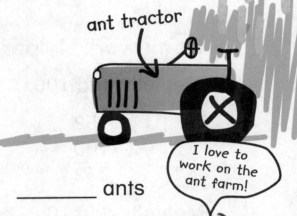

ant tractor

I love to work on the ant farm!

$\underline{\hspace{2cm}}$ ants

12. Ben has 115 pennies. His dad gives him more pennies. Now Ben has 215 pennies. How many pennies did his dad give him?

$\underline{\hspace{2cm}}$ pennies

Test Practice

13. Find the sum.

$$825 + 100 = \underline{\hspace{1.5cm}}$$

725   900   925   920

◯   ◯   ◯   ◯

Math at Home Practice saying a number and having your child add 10 or 100 to the number mentally.

Check My Progress

Vocabulary Check

Complete each sentence.

| tens | regroup | ones |

1. In the number 352, the 5 is in the _____ place.

2. In the number 352, the 2 is in the _____ place.

3. If the ones add up to more than 9, you must _____.

Concept Check

Make a hundred to add.

4.   25 + 197

_____ + _____ + 197

22 + _____ = _____

So, 25 + 197 = _____.

5.   314 + 96

_____ + _____ + 96

310 + _____ = _____

So, 314 + 96 = _____.

Add.

6. $200 + 100 =$ _____

7. $400 + 200 =$ _____

8. $\begin{array}{r} 100 \\ + \ 300 \\ \hline \end{array}$

9. $\begin{array}{r} 700 \\ + \ 100 \\ \hline \end{array}$

10. $\begin{array}{r} 500 \\ + \ 400 \\ \hline \end{array}$

11. $\begin{array}{r} 836 \\ + \ 100 \\ \hline \end{array}$

12. $\begin{array}{r} 390 \\ + \ 100 \\ \hline \end{array}$

13. $\begin{array}{r} 567 \\ + \ 10 \\ \hline \end{array}$

14. $564 + 10 =$ _____

15. $626 + 100 =$ _____

Find the missing addend.

16. $239 +$ _____ $= 249$

17. $853 +$ _____ $= 953$

Test Practice

18. Sophia counts 297 watermelon seeds. Isabella counts 100 more seeds than Sophia. How many seeds does Isabella count?

   100   197   297   397

   ◯   ◯   ◯   ◯

Name _____

Regroup Ones to Add

Lesson 4

ESSENTIAL QUESTION
How can I add three-digit numbers?

Explore and Explain

Watch Tools

Woof!

_____ + _____ = _____ books

 Teacher Directions: Model using base-ten blocks. Our library has 146 books about dogs. It has 145 books about cats. How many books about cats and dogs does the library have? Write the number sentence.

See and Show

Processes &Practices

Find 135 + 328.

Step 1 Add the ones. If there are 10 or more ones, regroup 10 ones as 1 ten. Write the 1 in the tens column.

hundreds	tens	ones
	1	
1	3	5
+ 3	2	8
		3

Step 2 Add the tens.

hundreds	tens	ones
	1	
1	3	5
+ 3	2	8
	6	3

Step 3 Add the hundreds.

hundreds	tens	ones
	1	
1	3	5
+ 3	2	8
4	6	3

So, 135 + 328 = __463__ .

Use Work Mat 7 and base-ten blocks. Add.

1.

hundreds	tens	ones
4	3	6
+ 2	4	5

2.

hundreds	tens	ones
1	2	7
+ 6	4	8

Talk Math   How is three-digit addition like two-digit addition?

On My Own

Add. Use base-ten blocks if needed.

3.
```
    4 │ 6 │ 8
  +   │ 2 │ 3
```

4.
```
  5 │ 6 │ 1
+   │ 2 │ 6
```

5.
```
    4 │ 1 │ 1
  + 3 │ 7 │ 9
```

6.
```
    236
  + 518
```

7.
```
    468
  +  18
```

8.
```
    427
  + 144
```

9.
```
    306
  + 408
```

10.
```
     28
  + 515
```

11.
```
    273
  + 224
```

12.
```
    749
  +   9
```

13.
```
    146
  + 253
```

14.
```
    135
  +  16
```

15.
```
    305
  + 306
```

16.
```
    607
  +  13
```

17.
```
    782
  + 103
```

Problem Solving

18. There are 247 red cars. There are 438 black cars. How many red and black cars in all?

_____ cars

19. Mr. Archer picked 416 apples last month. He picked 336 apples this month. How many apples did he pick in all?

I can't wait to meet the new family on the block!

_____ apples

20. The park ranger counted 105 black bears last year. This year, he counts 128 black bears. How many black bears did he count in all?

_____ black bears

 Write Math Explain how to regroup ones to make a ten when adding three-digit numbers.

_____

_____

_____

My Homework

Homework Helper Need help? connectED.mcgraw-hill.com

Find 225 + 137.

Step 1 Add the ones.
Regroup 10 ones
as 1 ten.
Step 2 Add the tens.
Step 3 Add the hundreds.

hundreds	tens	ones
	1	
2	2	5
+ 1	3	7
3	6	2

Practice

Add.

1.

hundreds	tens	ones
	☐	
4	6	3
+ 1	1	8

2.

hundreds	tens	ones
	☐	
1	8	2
+ 1	0	9

3.

hundreds	tens	ones
	☐	
3	3	4
+ 1	2	8

4.

	☐	
7	4	4
+ 1	3	8

5.

	☐	
3	6	3
+ 2	1	9

6.

	☐	
8	2	7
+ 1	5	5

Add.

7. 335
   + 116

8. 135
   + 219

9. 425
   + 148

10. 149 children are on a playground. 131 more children come to the playground. How many children are on the playground now?

_____ children

11. There are 152 tadpoles and 138 frogs in a pond. How many frogs and tadpoles are there in all?

The pond in our town rules!

_____ frogs and tadpoles

Test Practice

12. Find the sum.

$$368 + 619 = \underline{\hspace{1cm}}$$

919 ○   968 ○   987 ○   997 ○

Math at Home Have your child explain how to regroup to add 215 + 215.

Name

Regroup Tens to Add

Explore and Explain

  Watch Tools

 STOP

_____ + _____ = _____

Teacher Directions: Model using base-ten blocks. My town has 163 stop signs. There are also 181 street lights. How many stop signs and street lights does my town have? Write the number sentence.

See and Show

Find 375 + 462.

Step 1 Add the ones.

hundreds	tens	ones
3	7	5
+ 4	6	2
		7

Step 2 Add the tens. If there are 10 or more tens, regroup 10 tens as 1 hundred. Write 1 in the hundreds column.

hundreds	tens	ones
1		
3	7	5
+ 4	6	2
	3	7

Step 3 Add the hundreds.

hundreds	tens	ones
1		
3	7	5
+ 4	6	2
8	3	7

So, 375 + 462 = __837__.

Use Work Mat 7 and base-ten blocks. Add.

1.

hundreds	tens	ones
□		
2	4	3
+ 3	8	5

2.

hundreds	tens	ones
□		
5	6	2
+ 1	7	4

Talk Math How is regrouping ones different from regrouping tens?

Name _____

On My Own

Add. Use base-ten blocks if needed.

3.
☐		
4	5	6
+ 2	9	1

4.
☐		
7	3	2
+ 1	6	7

5.
☐		
3	2	4
+	9	3

6.
☐		
6	8	4
+	2	4

7.
☐		
4	8	5
+ 3	3	2

8.
☐		
8	9	5
+	1	1

9.
```
  363
+ 281
```

10.
```
  286
+ 121
```

11.
```
  384
+ 134
```

12.
```
  466
+  60
```

13.
```
  352
+ 475
```

14.
```
  382
+  51
```

15.
```
  601
+ 281
```

16.
```
  558
+ 360
```

17.
```
  387
+ 122
```

Problem Solving

18. There are 156 men and 163 women at the town meeting. How many people came to the meeting?

_____ people

19. A forest ranger plants 132 pine trees and 191 oak trees. How many trees did the ranger plant in all?

Finally, a chance to rest!

_____ trees

20. There are 124 red park benches and 185 blue park benches. How many red and blue park benches are there in all?

_____ benches

Write Math How do you know when to regroup?

_____

_____

_____

Name

My Homework

Homework Helper

Need help? connectED.mcgraw-hill.com

Find 353 + 176.

Step 1 Add the ones.

Step 2 Add the tens. If there are 10 or more tens, regroup 10 tens as 1 hundred.

Step 3 Add the hundreds.

So, 353 + 176 = 529.

	hundreds	tens	ones
	☐ 1		
	3	5	3
+	1	7	6
	5	2	9

Practice

Add.

1.

	hundreds	tens	ones
	☐		
	3	7	5
+	4	5	4

2.

	hundreds	tens	ones
	☐		
	2	5	7
+	6	9	1

3.

	hundreds	tens	ones
	☐		
	3	2	5
+	1	9	2

4.

	☐		
	1	9	8
+	1	2	1

5.

	☐		
	6	5	6
+	1	5	3

6.

	☐		
	1	8	5
+	2	9	4

Add.

7. 194
    + 333
    ___

8. 352
    + 281
    ___

9. 736
    + 170
    ___

Yay!
A sunny day!

10. Last year there were 252 sunny days.
    This year there were 164 sunny days.
    How many sunny days were there in all?

    _____ days

11. Logan has 156 baseball cards. Caden has
    182 baseball cards. How many baseball
    cards do they have in all?

    _____ cards

Test Practice

12. Find the sum.

    $373 + 465 =$ _____

    738   838   938   818
    ○   ○   ○   ○

Math at Home Have your child explain how to regroup tens to solve $185 + 292$.

382   Chapter 6 • Lesson 5

Add Three-Digit Numbers

Explore and Explain

I'm crunchy!

_____ + _____ = _____ people

 Teacher Directions: Model using base-ten blocks. On Friday, 347 people went to the movies. On Saturday, 485 people went to the movies. How many people went to the movies on these two days in all? Write the number sentence.

See and Show

Find 582 + 268.

Step 1 Add the ones, regroup.

Step 2 Add the tens, regroup.

Step 3 Add the hundreds.

So, 582 + 268 = __850__.

hundreds	tens	ones
1	1	
5	8	2
+ 2	6	8
8	5	0

Use Work Mat 7 and base-ten blocks. Add.

1.

hundreds	tens	ones
☐	☐	
3	5	7
+ 4	5	4

2.

hundreds	tens	ones
☐	☐	
2	6	1
+	6	9

3.

hundreds	tens	ones
☐	☐	
1	1	8
+ 1	9	9

4.

hundreds	tens	ones
☐	☐	
5	2	3
+ 2	8	7

Talk Math Explain how you solved Exercise 2.

Name _____

On My Own

Add. Use base-ten blocks if needed.

5.
$$\begin{array}{r} \square\ \square \\ 1\ 6\ 6 \\ +\ 1\ 4\ 7 \\ \hline \end{array}$$

6.
$$\begin{array}{r} \square\ \square \\ 7\ 2\ 0 \\ +\ 1\ 9\ 5 \\ \hline \end{array}$$

7.
$$\begin{array}{r} \square\ \square \\ 4\ 5\ 8 \\ +\ \ 6\ 6 \\ \hline \end{array}$$

8.
$$\begin{array}{r} 34 \\ +\ 515 \\ \hline \end{array}$$

9.
$$\begin{array}{r} 632 \\ +\ 299 \\ \hline \end{array}$$

10.
$$\begin{array}{r} 327 \\ +\ 127 \\ \hline \end{array}$$

11.
$$\begin{array}{r} 129 \\ +\ 386 \\ \hline \end{array}$$

12.
$$\begin{array}{r} 103 \\ +\ 609 \\ \hline \end{array}$$

13.
$$\begin{array}{r} 123 \\ +\ 785 \\ \hline \end{array}$$

14.
$$\begin{array}{r} 843 \\ +\ \ 77 \\ \hline \end{array}$$

15.
$$\begin{array}{r} 422 \\ +\ 309 \\ \hline \end{array}$$

16.
$$\begin{array}{r} 215 \\ +\ 193 \\ \hline \end{array}$$

17.
$$\begin{array}{r} 395 \\ +\ 238 \\ \hline \end{array}$$

18.
$$\begin{array}{r} 725 \\ +\ 174 \\ \hline \end{array}$$

19.
$$\begin{array}{r} 662 \\ +\ 199 \\ \hline \end{array}$$

Problem Solving

20. One day Keith skipped 145 rocks over the river. The next day he skipped 165 rocks. How many rocks did he skip altogether?

_____ rocks

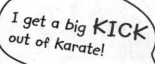
I get a big **KICK** out of karate!

21. One karate class has 156 students. The other karate class has 176 students. How many students are in both karate classes?

_____ students

22. 389 people went to the zoo on Saturday. 592 people went to the zoo on Sunday. How many people went to the zoo in all?

_____ people

Write Math Explain how adding three-digit numbers is different than adding two-digit numbers.

_____

_____

_____

Name _____

My Homework

Homework Helper   Need help? connectED.mcgraw-hill.com

Find 536 + 189.

Step 1 There are 10 or more ones, regroup.

Step 2 Add tens. There are 10 or more tens, regroup.

Step 3 Add hundreds.
So, 536 + 189 = 725.

hundreds	tens	ones
1	1	
5	3	6
+ 1	8	9
7	2	5

Practice

Add.

1.

hundreds	tens	ones
2	5	6
+ 4	5	6

2.

hundreds	tens	ones
4	6	4
+ 1	8	5

3.

hundreds	tens	ones
3	5	9
+ 1	6	7

4.

5	7	3
+ 3	6	3

5.

3	6	7
+ 1	5	5

6.

4	7	7
+ 2	3	4

Add.

7.   285
   + 229

8.   476
   + 345

9.   394
   + 217

10. There are 189 animals at one zoo. There are 158 animals at another zoo. How many animals are there at both zoos?

_____ animals

11. Ella has 199 marbles. Kaitlyn has 137 marbles. How many marbles do the girls have in all?

 Don't lose your marbles. You can do it!

_____ marbles

Test Practice

12. Find the sum.

$$369 + 264 = _____$$

633 ◯   533 ◯   433 ◯   133 ◯

 Math at Home Give your child a three-digit addition problem. Have him or her show you how he or she could solve the problem and then write the answer.

Name _____

Rewrite Three-Digit Addition

Explore and Explain

143 + 167

hundreds	tens	ones
+		

Be proud of your town!

_____ firefighters and police officers

Teacher Directions: There are 143 firefighters and 167 police officers in my town. How many firefighters and police officers are in my town in all? Write the number in the place-value chart and add.

See and Show

You can rewrite a problem to add.

Find 366 + 264.

Step 1 Write one addend below the other addend. Line up the ones, tens, and hundreds.

Step 2 Add. Regroup if necessary.

hundreds	tens	ones
☐	☐	
3	6	6
+ 2	6	4
6	3	0

Rewrite the problems. Add.

1. 634 + 125

2. 245 + 236

3. 743 + 163

4. 264 + 566

5. 396 + 378

6. 237 + 592

Talk Math How is rewriting three-digit addition different than rewriting two-digit addition?

Name

On My Own

Rewrite the problems. Add.

7. 363 + 278

+

8. 236 + 265

+

9. 298 + 516

+

10. 373 + 495

+

11. 195 + 645

+

12. 165 + 454

+

13. 652 + 258

+

14. 268 + 195

+

15. 496 + 349

+

Problem Solving

16. 395 ants crawled around the ant hill.
111 more ants came over to the ant hill.
How many ants are there now?

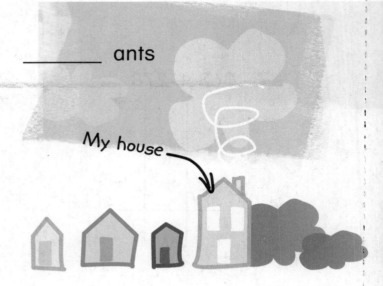

_____ ants

17. There are 153 yellow houses
and 168 blue houses in
the neighborhood. How many
yellow and blue houses
are there in all?

My house

_____ yellow and
blue houses

18. Last week, Amelia sold 193 cups of lemonade.
This week, she sold 129 cups. How many cups
of lemonade has she sold in both weeks?

_____ cups

Explain how you rewrite three-digit
addition problems to solve.

_____

_____

_____

Name

My Homework

Homework Helper Need help? connectED.mcgraw-hill.com

Find 295 + 185.

Step 1 Rewrite.

Step 2 Add. Regroup if necessary.

$$
\begin{array}{r}
\boxed{1}\ \boxed{1} \\
2\ 9\ 5 \\
+\ 1\ 8\ 5 \\
\hline
4\ 8\ 0
\end{array}
$$

Helpful Hint
Line up the ones, tens, and hundreds.

Practice

Rewrite the numbers. Add.

1. 654 + 286

2. 518 + 229

3. 846 + 119

4. 556 + 148

+ _____

5. 685 + 295

+ _____

6. 584 + 371

+ _____

Rewrite the problems. Add.

7. $172 + 217$

+ _____

8. $362 + 264$

+ _____

9. $239 + 375$

+ _____

10. There are 129 boats in the lake. There are 197 jet skis in the lake. How many boats and jet skis are in the lake?

Weeeeee!

_____ boats and jet skis

Test Practice

11. Mark the answer that shows how to rewrite and solve the number sentence.

$$427 + 228 = _____$$

$$\begin{array}{r} 427 \\ + 228 \\ \hline 6415 \end{array}$$

○

$$\begin{array}{r} {\scriptstyle 3\,11\,17} \\ \cancel{427} \\ - 228 \\ \hline 199 \end{array}$$

○

$$\begin{array}{r} 427 \\ + 228 \\ \hline 645 \end{array}$$

○

$$\begin{array}{r} {\scriptstyle 1} \\ 427 \\ + 228 \\ \hline 655 \end{array}$$

○

Math at Home Write $374 + 185$ on a piece of paper. Have your child rewrite the problem and solve it.

Name

Problem Solving
STRATEGY: Guess, Check, and Revise

They just dropped me off!

A store has three bins of potatoes with 183, 127, and 117 potatoes in separate bins. What two bins have the sum of 300 potatoes?

1 Understand Underline what you know.
Circle what you need to find.

2 Plan How will I solve the problem?

3 Solve Guess, check, and revise.
Guess: 183 and 127
Check: 183 + 127 = 310
Revise: 310 is too high; guess again.
Guess: 183 and 117
Check: 183 + 117 = 300

So, the bins with __183__ and __117__ have 300 potatoes.

4 Check Is my answer reasonable? Explain.

Practice the Strategy

A mail carrier delivered 430 letters last month. This month he delivered less letters than last month. He delivered 850 letters in both months in all. How many letters did he deliver this month?

1 Understand <u>Underline</u> what you know.
(Circle) what you need to find.

2 Plan How will I solve the problem?

3 Solve I will...

4 Check Is my answer reasonable? Explain.

Apply the Strategy

1. A store sells raisins in packages of 130, 165,
    and 170. Rosa wants to buy 300 raisins.
    Which two packages will she buy?

    packages of _____ and _____

2. Alexa's ant farm has 120 ants. She buys
    another ant farm. She now has 300 ants.
    How many ants are in the second ant farm?

    Guys, look
    what I found!

    _____ ants

3. Tao needs to read three books that total
    500 pages. Which three books should he read?

Book	Pages
1	115
2	150
3	200
4	185

    Books _____, _____, and _____

Review the Strategies

Choose a strategy
- Guess, check, and revise.
- Act it out.
- Use logical reasoning.

4. Clara counted 185 flowers on her walk. Her dad counted 139 flowers. How many flowers did they count altogether?

_____ flowers

5. Chelsea collects sand dollars. The number she has collected is greater than 400 and less than 460. The number in the hundreds place is one less than the number in the tens place. The number in the ones place is two more than the number in the tens place. How many sand dollars has Chelsea collected?

_____ sand dollars

6. One pet store has 289 angel fish. Another pet store has 132 angel fish. How many angel fish do the stores have altogether?

_____ angel fish

Name _____

My Homework

163 girls went to see fireworks.
About 300 people went to see the
fireworks in all. How many boys
went to see the fireworks?

eHelp

1 Understand Underline what you know.
Circle what you need to find.

2 Plan How will I solve the problem?

3 Solve Guess, check and revise.

Guess: 147 boys
Check: $163 + 147 = 310$
Revise: 147 is too high; guess a lower number.
Guess: 137
Check: $163 + 137 = 300$
So, 137 boys went to the fireworks.

4 Check Is my answer reasonable?

Underline what you know. Circle what you need to find out. Guess, check, and revise to solve.

1. 350 people went to the movie on Saturday. Fewer people went to the movie on Sunday. About 600 people went to the movies in all. How many people went to the movie on Sunday?

_____ people

2. Pam has 65 purple flowers, 50 yellow flowers, and 75 red flowers in her garden. She needs a total of 125 flowers in 2 colors. Which color of flowers should she pick?

_____ flowers

3. My garden has 145 corn stalks. It also has some tomato plants. There are about 245 corn and tomato plants in my garden. About how many tomato plants are in my garden?

I love corny jokes!

_____ tomato plants

Name _____

My Review

Vocabulary Check

ones   hundreds   tens   regroup   place value

Complete each sentence.

1. The numbers 0–9 are _____.

2. The value given to a digit by its place
   in a number is its _____.

3. The numbers 10–99 are _____.

4. The numbers 100–999 are _____.

5. When you _____, you take
   a number apart and write it in a new way.

Concept Check

Add.

6.   200
   + 100
   _____

7.   400
   + 500
   _____

8.   500
   + 300
   _____

Add.

9. 354
    + 100
    _____

10. 327
    + 10
    _____

11. 356
    + 100
    _____

12. 275
    + 116
    _____

13. 434
    + 247
    _____

14. 367
    + 226
    _____

15. 777
    + 133
    _____

16. 397
    + 159
    _____

17. 487
    + 247
    _____

Rewrite the problems. Add.

18. 395 + 154

    +
    _____

19. 225 + 353

    +
    _____

20. 348 + 273

    +
    _____

Name _____

Problem Solving

21. Nathan collected 109 pine cones. Elizabeth collected 100 pine cones. How many pine cones did they collect in all?

Thanks!

You smell great!

_____ pine cones

22. Olivia bakes 154 cookies with her grandma. Connor bakes 146 cookies with his grandma. How many cookies do Olivia and Connor bake in all?

_____ cookies

Test Practice

23. Last month, Riley watched 115 hours of television. This month, she will watch fewer hours of television. She wants to watch 200 hours for both months. How many hours can Riley watch this month?

85   95   185   315

○   ○   ○   ○

Show the ways you can add three-digit numbers.

Regroup ones.

```
    ☐
    1 0 1
+   1 2 9
─────────
```

Regroup tens.

```
    ☐
    2 3 4
+   3 8 3
─────────
```

ESSENTIAL QUESTION

How do I add three-digit numbers?

Regroup tens and ones.

```
  ☐ ☐
    3 6 4
+   4 5 8
─────────
```

Add mentally.

$358 + 100 = $ _____

$782 +$ _____ $= 792$

MAIN ST.

LITTLE G.

Remember to visit me!

You will go far!

Chapter 7 Subtract Three-Digit Numbers

ESSENTIAL QUESTION

How can I subtract three-digit numbers?

My School is Cool!

Watch a video!

Watch

Chapter 7 Project

Reference Posters for Subtracting Three-Digit Numbers

1. Work together to determine how your group will show subtracting three-digit numbers on your poster.
2. Your plans should include a list of steps for strategies and drawings of concepts.
3. Plan an example of your poster in the space below.

Name

Subtract.

1.   9
   − 6

2.   8
   − 8

3.   17
   − 8

4.   14
   − 5

5.   60
   − 30

6.   87
   − 53

7.   62
   − 37

8.   72
   − 25

9.   95
   − 14

10.   90
    − 60

11.   88
    − 61

12.   43
    − 25

13. Aiden sees eight squirrels.
Two run away. How many
squirrels does Aiden see now?

_____ squirrels

How Did I Do?

Shade the boxes to show the
problems you answered correctly.

| 1 | 2 | 3 | 4 | 5 | 6 | 7 | 8 | 9 | 10 | 11 | 12 | 13 |

My Math Words

Review Vocabulary

| difference | ones | regroup | subtract | tens |

Write a review word that has the same meaning as the underlined words or numbers.

I will <u>take away</u> numbers.

_____

<u>5</u>7 – <u>3</u>9

_____

23 – 1<u>2</u>

_____

I will <u>take apart</u> the ones.

_____

The <u>answer</u> is 11.

_____

The <u>answer</u> is 18.

_____

My Vocabulary Cards

Vocab

Directions:
Ideas for Use

- Ask students to use the blank cards to draw or write words that will help them with concepts like three-digit subtraction or subtract across zeros.

- Have students use the blank cards to write basic subtraction facts. They should write the answer on the back of each card.

My Foldable

Name _____

Take Apart Hundreds to Subtract

Lesson 1

ESSENTIAL QUESTION
How can I subtract three-digit numbers?

Explore and Explain

Stick together. We don't want to get lost!

_____ crayons

Teacher Directions: At the beginning of the year, Mrs. Jones put 403 new crayons in the crayon bucket. After the first month of school, 97 crayons were lost. How many crayons were left in the bucket? Use base-ten blocks to help you solve the problem. Draw the blocks you used and write the answer.

See and Show

You can take apart hundreds to subtract mentally.

Take apart 800 as 700 and 100 since it is easier to subtract 98 from 100.

$$800 - 98$$

700   100

$100 - 98 =$ __2__

$700 + 2 =$ __702__

So, $800 - 98 =$ __702__.

Take apart 401 as 301 and 100 since it is easier to subtract 97 from 100.

$$401 - 97$$

Helpful Hint
What hundreds do you need to use?

301   100

$100 - 97 =$ __3__

$301 + 3 =$ __304__

So, $401 - 97 =$ __304__.

Take apart hundreds to subtract.

1. $300 - 99$

_____ _____

_____ $- 99 =$ _____

_____ $+$ _____ $=$ _____

So, $300 - 99 =$ _____.

2. $422 - 98$

_____ _____

_____ $- 98 =$ _____

_____ $+$ _____ $=$ _____

So, $422 - 98 =$ _____.

Talk Math   In Exercise 2, why is 2 added back?

Name

On My Own

Take apart hundreds to subtract.

3.   240 − 99

_____   _____

_____ − 99 = _____

_____ + _____ = _____

So, 240 − 99 = _____.

4.   700 − 98

_____   _____

_____ − 98 = _____

_____ + _____ = _____

So, 700 − 98 = _____.

5.   542 − 97

_____   _____

_____ − 97 = _____

_____ + _____ = _____

So, 542 − 97 = _____.

6.   702 − 98

_____   _____

_____ − 98 = _____

_____ + _____ = _____

So, 702 − 98 = _____.

7.   200 − 97

_____   _____

_____ − 97 = _____

_____ + _____ = _____

So, 200 − 97 = _____.

8.   711 − 99

_____   _____

_____ − 99 = _____

_____ + _____ = _____

So, 711 − 99 = _____.

Problem Solving

9. Eli counts 335 sunflower seeds. He places 98 of the seeds in a bowl. How many seeds were left?

_____ sunflower seeds

10. Rachel has 148 blueberries. She gives 97 blueberries to her brother. How many blueberries does Rachel have left?

_____ blueberries

11. 382 people were in our school auditorium. 99 people left. How many people are still in the auditorium?

_____ people

HOT Problem Explain in words how to solve 700 − 97.

_____

_____

_____

Name
....

My Homework

Homework Helper 🏠 **Need help?** 👆 connectED.mcgraw-hill.com

You can take apart hundreds
to mentally subtract $324 - 99$.

Take apart 324 as 224 and 100.

It is easier to subtract 99
from 100.

$324 - 99$

224   100

$100 - 99 = 1$

$224 + 1 = 225$

So, $324 - 99 = 225$.

Practice

Take apart hundreds to subtract.

1. $835 - 98$

_____ _ _____

_____ _ _____ = _____

_____ + _____ = _____

So, $835 - 98 =$ _____.

2. $748 - 97$

_____ _ _____

_____ _ _____ = _____

_____ + _____ = _____

So, $748 - 97 =$ _____.

Take apart hundreds to subtract.

3.   395 – 99

_____ _____

_____ – _____ = _____

_____ + _____ = _____

So, 395 – 99 = _____.

4.   600 – 97

_____ _____

_____ – _____ = _____

_____ + _____ = _____

So, 600 – 97 = _____.

5. Students go to school for 180 days. 96 school days were sunny. How many school days were not sunny?

_____ days

It's definitely "not sunny" today.

Test Practice

6. How would you take apart 355 to solve 355 – 94?

    355 and 100   ◯   300 and 55   ◯

    255 and 100   ◯   255 and 94   ◯

Math at Home Have your child solve 321 – 99 by taking apart hundreds to subtract.

Name

Subtract Hundreds

Explore and Explain

What goes up, must come down!

_____ steps

 Teacher Directions: There are 400 steps in my school. My class decided to walk up all of the steps. We have walked up 300 steps already. How many steps do we have left? Use base-ten blocks. Draw the blocks. Write how many steps are left.

Online Content at connectED.mcgraw-hill.com

Copyright © The McGraw-Hill Companies, Inc.   Andersen Ross/Blend Images/Getty Images

See and Show

You can use basic subtraction facts to help subtract hundreds.

Find 500 − 300.

Mind if I subtract a few of these carrots?

$$
\begin{array}{r}
5 \text{ hundreds} \\
- \ 3 \text{ hundreds} \\
\hline
2 \text{ hundreds}
\end{array}
$$

500 − 300 = __200__

Subtract.

1. 800 − 100 = _____

2. 200 − 200 = _____

3. $\begin{array}{r} 700 \\ -\ 100 \\ \hline \end{array}$

4. $\begin{array}{r} 400 \\ -\ 200 \\ \hline \end{array}$

5. $\begin{array}{r} 500 \\ -\ 100 \\ \hline \end{array}$

6. $\begin{array}{r} 700 \\ -\ 300 \\ \hline \end{array}$

Talk Math What subtraction fact can you use to find 900 − 800?

Name _____

Helpful Hint
Use the subtraction facts you know to subtract hundreds.

On My Own

Subtract.

7. $800 - 300 =$ _____

8. $600 - 600 =$ _____

9. $300 - 100 =$ _____

10. $900 - 200 =$ _____

11. $800 - 400 =$ _____

12. $700 - 500 =$ _____

13. $200 - 200 =$ _____

14. $400 - 100 =$ _____

15.
$$\begin{array}{r} 700 \\ -\ 400 \\ \hline \end{array}$$

16.
$$\begin{array}{r} 800 \\ -\ 200 \\ \hline \end{array}$$

17.
$$\begin{array}{r} 300 \\ -\ 300 \\ \hline \end{array}$$

18.
$$\begin{array}{r} 500 \\ -\ 400 \\ \hline \end{array}$$

19.
$$\begin{array}{r} 500 \\ -\ 300 \\ \hline \end{array}$$

20.
$$\begin{array}{r} 200 \\ -\ 100 \\ \hline \end{array}$$

21.
$$\begin{array}{r} 600 \\ -\ 200 \\ \hline \end{array}$$

22.
$$\begin{array}{r} 900 \\ -\ 700 \\ \hline \end{array}$$

23.
$$\begin{array}{r} 700 \\ -\ 200 \\ \hline \end{array}$$

24.
$$\begin{array}{r} 400 \\ -\ 300 \\ \hline \end{array}$$

25.
$$\begin{array}{r} 800 \\ -\ \ \ 0 \\ \hline \end{array}$$

26.
$$\begin{array}{r} 800 \\ -\ 800 \\ \hline \end{array}$$

27. 700 people came to my school's spring concert. 500 people came to our winter concert. How many more people came to the spring concert?

_____ people

28. My school made flags for Memorial Day. There were 600 flags. 200 students took their flags home. How many flags are left?

Stars and stripes FOREVER!

_____ flags

29. The floor in my classroom has 200 tiles. 100 tiles are blue. The rest of the tiles are white. How many white tiles are there?

_____ tiles

 Write Math   How is subtracting hundreds like subtracting ones?

_____

_____

_____

Name _____

My Homework

Homework Helper Need help? connectED.mcgraw-hill.com

Find 300 − 200.

$$\begin{array}{r} 3 \text{ hundreds} \\ - \ 2 \text{ hundreds} \\ \hline 1 \text{ hundred} \end{array}$$ $300 - 200 = 100$

Helpful Hint
You know 3 − 2 = 1.
So, 300 − 200 = 100.

Practice

Subtract.

1. 700 − 300 = _____

2. 500 − 200 = _____

3. $\begin{array}{r} 300 \\ - \ 100 \\ \hline \end{array}$

4. $\begin{array}{r} 100 \\ - \ 100 \\ \hline \end{array}$

5. $\begin{array}{r} 600 \\ - \ 400 \\ \hline \end{array}$

6. $\begin{array}{r} 800 \\ - \ 700 \\ \hline \end{array}$

7. $\begin{array}{r} 900 \\ - \ 100 \\ \hline \end{array}$

8. $\begin{array}{r} 500 \\ - \ 100 \\ \hline \end{array}$

9. $\begin{array}{r} 400 \\ - \ 400 \\ \hline \end{array}$

10. $\begin{array}{r} 400 \\ - \ 200 \\ \hline \end{array}$

Solve each word problem.

11. 800 people went to watch the ballet. 200 people left early. How many people stayed at the ballet?

_____ people

12. 500 boys and girls went to the basketball game. 300 girls were at the game. How many boys were at the game?

_____ boys

13. The cafeteria had pizza slices and hot dogs for lunch. 900 lunches were sold. 300 people bought hot dogs. How many pizza slices were sold?

_____ pizza slices

Test Practice

14. Which number sentence could help you solve 700 − 500?

7 − 2   2 + 5   5 + 2   7 − 5
○   ○   ○   ○

Math at Home Ask your child what number is 100 less than 500.

Mentally Subtract 10 or 100

Explore and Explain

Recess rocks!

_____ students

Teacher Directions: 135 students are on the playground. 100 students go inside. How many students are left on the playground? Use base-ten blocks to solve. Write the number.

See and Show

To subtract 100 or 10, think of facts you know.
Mentally subtract 100.

$$422 - 100$$

Helpful Hint
You know that 4 – 1 = 3.

$$422 - 100 = 322$$

I'm mentally subtracting my lunch from my lunchbox!

Mentally subtract 10.

$$460 - 10$$

You know that 6 – 1 = 5.

$$460 - 10 = 450$$

Subtract.

1.
$$329 - 100$$

2.
$$820 - 10$$

3.
$$363 - 100$$

4.
$$678 - 10$$

5.
$$724 - 10$$

6.
$$164 - 10$$

7. 900 – 100 = _____

8. 743 – 10 = _____

Talk Math Tell how to mentally subtract 10 or 100.

Name _____

On My Own

Subtract.

9. 491
    −100
    _____

10. 942
    − 10
    _____

11. 770
    − 10
    _____

12. 672
    −100
    _____

13. 853
    −100
    _____

14. 269
    − 10
    _____

15. 368
    − 10
    _____

16. 374
    −100
    _____

17. 982
    −100
    _____

18. 498 − 100 = _____

19. 533 − 100 = _____

Find the missing number.

20. 434 − _____ = 424

21. 371 − _____ = 271

22. 738 − _____ = 638

23. 270 − _____ = 260

Problem Solving

24. 863 flowers bloomed on the school playground. 100 of the flowers were eaten by bugs. How many flowers are on the playground now?

_____ flowers

25. 299 people came to the school play. 10 people had to leave early. How many people were still at the play?

The play must go on!

_____ people

26. 423 wildflowers grew in the field. Embry picked 10 of them. How many wildflowers are left in the field?

_____ wildflowers

Write Math   Explain how you would mentally subtract 10 from 900.

_____

_____

_____

Name _____

My Homework

Homework Helper   **Need help?** connectED.mcgraw-hill.com

Mentally subtract 100.

$$\begin{array}{r} 567 \\ -\ 100 \\ \hline 467 \end{array}$$

Mentally subtract 10.

$$\begin{array}{r} 567 \\ -\ 10 \\ \hline 557 \end{array}$$

Helpful Hint
You know that
$5 - 1 = 4$.

You know that
$6 - 1 = 5$.

Practice

Subtract.

1. $\begin{array}{r} 477 \\ -\ 100 \\ \hline \end{array}$	2. $\begin{array}{r} 378 \\ -\ 10 \\ \hline \end{array}$	3. $\begin{array}{r} 879 \\ -\ 100 \\ \hline \end{array}$
4. $\begin{array}{r} 245 \\ -\ 10 \\ \hline \end{array}$	5. $\begin{array}{r} 849 \\ -\ 100 \\ \hline \end{array}$	6. $\begin{array}{r} 320 \\ -\ 10 \\ \hline \end{array}$

Find the missing numbers.

7. 358 − _____ = 258

8. 843 − _____ = 833

9. 954 − _____ = 944

10. 700 − _____ = 600

11. Joan counts 143 birds by the pond. 100 birds fly away. How many birds are left by the pond?

Where's your tail?

_____ birds

12. There were 694 tadpoles in the lake. 100 of the tadpoles turned into frogs. How many tadpoles were left in the lake?

_____ tadpoles

Test Practice

13. Find 363 − 100.

163 ⃝   263 ⃝   363 ⃝   463 ⃝

Math at Home Practice saying a number and having your child subtract 10 or 100 from the number mentally.

430   Chapter 7 • Lesson 3

Name _____

Vocabulary Check

Draw lines to match.

1. **difference**   To find the difference between two sets.

2. **subtract**   To take apart a number to write it in a new way.

3. **regroup**   The answer to a subtraction problem.

Concept Check

Take apart addends to subtract.

4.   345 − 98

_____ _____

_____ − _____ = _____

_____ + _____ = _____

So, 345 − 98 = _____.

5.   926 − 99

_____ _____

_____ − _____ = _____

_____ + _____ = _____

So, 926 − 99 = _____.

Subtract.

6. 900 − 400 = _____   **7.** 700 − 300 = _____

8.   393
        − 100
      _____

9.   264
        − 10
      _____

10.   737
        − 100
      _____

Find the missing number.

11. 394 − _____ = 294   **12.** 842 − _____ = 832

13. 535 − _____ = 435   **14.** 253 − _____ = 243

15. Brooklyn read 125 books over the summer.
Aubrey read 10 less books than Brooklyn.
How many books did Aubrey read?

_____ books

Test Practice

16. Find 300 − 300.

600   300   100   0
○   ○   ○   ○

Name _____

Regroup Tens

Explore and Explain

Watch ▶   Tools

Only 360 more and I break the school record!

Go!

_____ jump ropes

Teacher Directions: Use base-ten blocks to solve. Mr. Hicks, our gym teacher, had 145 jump ropes. 36 jump ropes broke this year. How many jump ropes does Mr. Hicks have left? Write the number.

See and Show

Find 652 − 429.

Step 1 Subtract the ones.
You cannot
subtract 9 from 2.
Regroup 1 ten as
10 ones.

hundreds	tens	ones
	4	12
6	5̶	2̶
− 4	2	9
		3

Step 2 Subtract the tens.

hundreds	tens	ones
	4	12
6	5̶	2̶
− 4	2	9
	2	3

Step 3 Subtract the
hundreds.

hundreds	tens	ones
	4	12
6	5̶	2̶
− 4	2	9
2	2	3

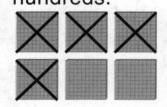

652 − 429 = __223__

Use Work Mat 7 and base-ten blocks. Subtract.

1.

hundreds	tens	ones
	☐	☐
5	4	6
−	1	7

2.

hundreds	tens	ones
	☐	☐
7	8	3
− 4	3	9

Talk Math How is subtracting three-digit numbers like subtracting two-digit numbers?

Name

On My Own

Use Work Mat 7 and base-ten blocks. Subtract.

3.

hundreds	tens	ones
	☐	☐
3	8	2
− 1	2	8

4.

hundreds	tens	ones
	☐	☐
4	6	7
−	4	9

5.

	☐	☐
5	7	5
−	6	6

6.

	☐	☐
8	6	3
− 2	1	8

7.

	☐	☐
2	6	4
− 1	3	5

8.
```
  754
− 507
```

9.
```
  455
− 326
```

10.
```
  930
− 428
```

11.
```
  780
− 436
```

12.
```
  652
−  35
```

13.
```
  931
−   6
```

14.
```
  387
−  18
```

15.
```
  423
− 119
```

16.
```
  540
−  15
```

Problem Solving

17. Ada jumped 382 times in a row. Helen jumped 277 times in a row. How many more jumps in a row did Ada do than Helen?

_____ jumps

18. A store had 472 bouncy balls. They sold 155. How many bouncy balls were left?

_____ bouncy balls

19. Lisa hit 294 golf balls in one day. She hit 149 in the morning. How many golf balls did she hit in the afternoon?

_____ golf balls

HOT Problem Andre wrote $381 - 165 = 224$. Tell why Andre is wrong. Make it right.

_____

_____

_____

Name _____

My Homework

Lesson 4

Regroup Tens

Homework Helper Need help? connectED.mcgraw-hill.com

Find 362 − 145.

Step 1 Subtract the ones. You cannot subtract 5 from 2. Regroup if needed.

Step 2 Subtract the tens.

Step 3 Subtract the hundreds.

hundreds	tens	ones
	5	12
3	6̶	2̶
− 1	4	5
2	1	7

Practice

Subtract.

1.

hundreds	tens	ones
	☐	☐
8	3	5
− 2	1	6

2.

hundreds	tens	ones
	☐	☐
9	5	2
− 6	3	7

3.

	☐	☐
1	5	3
−	4	4

4.

	☐	☐
6	4	2
− 2	1	8

5.

	☐	☐
7	3	4
− 2	2	8

Subtract.

6.
$$
\begin{array}{r}
153 \\
- 47 \\
\hline
\end{array}
$$

7.
$$
\begin{array}{r}
642 \\
- 215 \\
\hline
\end{array}
$$

8.
$$
\begin{array}{r}
754 \\
- 225 \\
\hline
\end{array}
$$

9.
$$
\begin{array}{r}
438 \\
- 129 \\
\hline
\end{array}
$$

10.
$$
\begin{array}{r}
362 \\
- 148 \\
\hline
\end{array}
$$

11.
$$
\begin{array}{r}
762 \\
- 349 \\
\hline
\end{array}
$$

12.
$$
\begin{array}{r}
647 \\
- 518 \\
\hline
\end{array}
$$

13.
$$
\begin{array}{r}
377 \\
- 163 \\
\hline
\end{array}
$$

14. 725 people came to my school's pancake breakfast. 318 of those people were children. How many adults were at the breakfast?

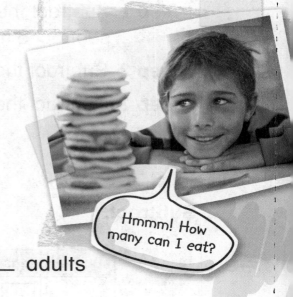

Hmmm! How many can I eat?

_____ adults

Test Practice

15. Which subtraction problem needs regrouping to solve?

392 − 222 ◯

385 − 266 ◯

692 − 321 ◯

295 − 172 ◯

 Math at Home Write a subtraction problem that requires regrouping tens on a piece of paper for your child to solve. Have your child explain each step to solving the problem. Try one-, two-, and three-digit subtraction problems.

Name _____

Regroup Hundreds

Explore and Explain

  Watch Tools

I am the most famous macaroni artist!

_____ pieces of macaroni

 Teacher Directions: Use base-ten blocks to solve. Gwen put 214 pieces of macaroni on her picture in art class. 120 pieces fell off. How many pieces of macaroni are still on the picture? Draw the blocks you used. Write the number.

See and Show

Find 539 − 285.
Step 1
Subtract the ones.

hundreds	tens	ones
5	3	9
− 2	8	5
		4

Step 2
Subtract the tens.
You cannot subtract
8 from 3. Regroup 1
hundred as 10 tens.

hundreds	tens	ones
4	13	9
− 5̸ 2	3̸ 8	5
	5	4

Step 3
Subtract the hundreds.

hundreds	tens	ones
4	13	9
− 5̸ 2	3̸ 8	5
2	5	4

So, 539 − 285 = __254__

Use Work Mat 7 and base-ten blocks. Subtract.

1.

hundreds	tens	ones
☐	☐	
5	6	8
− 1	9	7

2.

hundreds	tens	ones
☐	☐	
6	2	8
− 4	4	2

Talk Math   How do you know when to regroup?

Name _____

On My Own

Use Work Mat 7 and base-ten blocks. Subtract.

3.

hundreds	tens	ones
☐	☐	
5	8	6
− 2	9	5

4.

hundreds	tens	ones
☐	☐	
6	3	8
−	4	3

5.

```
  ☐   ☐
  4   5   9
  −   6   9
```

6.

```
  ☐   ☐
  7   3   9
− 5   4   1
```

7.

```
  ☐   ☐
  8   2   7
− 2   4   7
```

8.
```
  638
−  36
```

9.
```
  232
− 170
```

10.
```
  948
− 472
```

11.
```
  565
− 272
```

12.
```
  640
−  50
```

13.
```
  729
− 135
```

14.
```
  225
− 133
```

15.
```
  485
− 194
```

16.
```
  529
− 395
```

17. Lucy went to basketball camp. 325 second graders and 234 first graders were there. How many more second graders were at camp?

_____ more second graders

18. 417 boys were at the museum. 286 girls were at the museum. How many more boys were at the museum than girls?

A macaroni artist masterpiece.

_____ more boys

19. 475 parents came to the school's fall party. 294 parents had to leave early. How many parents are still at the fall party?

_____ parents

 Explain how to regroup hundreds.

_____

_____

_____

Name

My Homework

Lesson 5

Regroup
Hundreds

Homework Helper Need help? connectED.mcgraw-hill.com

Find 436 − 245.

Step 1 Subtract ones.

Step 2 Subtract the tens. You cannot
subtract 4 from 3. Regroup
1 hundred as 10 tens.

Step 3 Subtract the hundreds.

hundreds	tens	ones
3 ~~4~~	13 ~~3~~	6
− 2	4	5
1	9	1

Practice

Subtract.

1.

hundreds	tens	ones
☐	☐	
3	6	2
− 2	7	1

2.

hundreds	tens	ones
☐	☐	
7	3	2
− 4	4	1

3.

☐	☐	
8	3	5
−	5	5

4.

☐	☐	
7	3	5
− 6	7	2

5.

☐	☐	
2	1	7
− 1	4	4

Chapter 7 • Lesson 5   443

Copyright © The McGraw-Hill Companies, Inc.

Subtract.

6.
$$\begin{array}{r} 453 \\ -62 \\ \hline \end{array}$$

7.
$$\begin{array}{r} 721 \\ -131 \\ \hline \end{array}$$

8.
$$\begin{array}{r} 745 \\ -552 \\ \hline \end{array}$$

9.
$$\begin{array}{r} 375 \\ -292 \\ \hline \end{array}$$

10.
$$\begin{array}{r} 423 \\ -282 \\ \hline \end{array}$$

11.
$$\begin{array}{r} 434 \\ -243 \\ \hline \end{array}$$

12.
$$\begin{array}{r} 625 \\ -462 \\ \hline \end{array}$$

13.
$$\begin{array}{r} 278 \\ -184 \\ \hline \end{array}$$

14. 835 cows are in the field. 251 cows go in the barn. How many cows are still in the field?

I wish I could go to school!

_____ cows

Test Practice

15. Which subtraction problem does not need regrouping to solve?

$363 - 148$ ○

$734 - 371$ ○

$357 - 147$ ○

$367 - 288$ ○

 Math at Home Have your child explain when you need to regroup to solve a problem.

Name ....

Subtract Three-Digit Numbers

Lesson 6

ESSENTIAL QUESTION
How can I subtract three-digit numbers?

Explore and Explain

MILK

Moo juice!

_____ students

 Teacher Directions: Use base-ten blocks to solve. 355 students are eating lunch in the cafeteria. 166 students packed their lunch. How many students bought their lunch? Draw the blocks. Write the number.

Online Content at ⌒connectED.mcgraw-hill.com

Chapter 7 • Lesson 6   445

See and Show

Find 634 − 159.

Step 1 Subtract the ones. You cannot subtract 9 ones from 4 ones. Regroup 1 ten as 10 ones.

hundreds	tens	ones
☐	2	14
6	3̶	4̶
− 1	5	9
		5

Step 2 Subtract the tens. You cannot subtract 5 tens from 2 tens. Regroup 1 hundred as 10 tens.

hundreds	tens	ones
5	12	14
6̶	3̶	4̶
− 1	5	9
	7	5

Step 3 Subtract the hundreds. Now you can subtract 1 hundred from 5 hundreds.

hundreds	tens	ones
5	12	14
6̶	3̶	4̶
− 1	5	9
4	7	5

634 − 159 = __475__

Talk Math

Explain what you write in the box above the ones and the box above the tens when you regroup tens and hundreds.

Name

On My Own

Use Work Mat 7 and base-ten blocks. Subtract.

1.
$$\begin{array}{r} \square\ \square\ \square \\ 4\ 3\ 1 \\ -\ 3\ 4\ 5 \end{array}$$

2.
$$\begin{array}{r} \square\ \square\ \square \\ 7\ 6\ 2 \\ -\ \ \ 8\ 8 \end{array}$$

3.
$$\begin{array}{r} \square\ \square\ \square \\ 9\ 5\ 3 \\ -\ 7\ 6\ 4 \end{array}$$

4.
$$\begin{array}{r} 702 \\ -\ 211 \end{array}$$

5.
$$\begin{array}{r} 884 \\ -\ 197 \end{array}$$

6.
$$\begin{array}{r} 632 \\ -\ 444 \end{array}$$

7.
$$\begin{array}{r} 485 \\ -\ 296 \end{array}$$

8.
$$\begin{array}{r} 357 \\ -\ 169 \end{array}$$

9.
$$\begin{array}{r} 625 \\ -\ 438 \end{array}$$

10.
$$\begin{array}{r} 590 \\ -\ 184 \end{array}$$

11.
$$\begin{array}{r} 718 \\ -\ 628 \end{array}$$

12.
$$\begin{array}{r} 394 \\ -\ 185 \end{array}$$

13.
$$\begin{array}{r} 561 \\ -\ 273 \end{array}$$

14.
$$\begin{array}{r} 934 \\ -\ 395 \end{array}$$

15.
$$\begin{array}{r} 533 \\ -\ 203 \end{array}$$

Problem Solving

16. The post office had 912 stamps on Friday.
By Saturday 189 stamps were left.
How many stamps were sold on Friday?

_____ stamps

17. Dakota had 200 plastic bugs. He lost
some bugs. He has 155 bugs left. How
many bugs did he lose?

_____ bugs

18. There are 350 pieces of popcorn.
Ayana and Gabby eat 177 pieces.
How many pieces are left?

_____ pieces

HOT Problem 228 students like red. 293 students
like blue. 154 students like green. How many more
students like red or green than blue? Explain.

_____

_____

_____

My Homework

Homework Helper Need help? connectED.mcgraw-hill.com

Sometimes when you subtract, you need to regroup the tens and hundreds.

Find 621 − 475.

Step 1 Subtract the ones. You cannot subtract 5 from 1. Regroup 1 ten as 10 ones.

Step 2 Subtract the tens. You cannot subtract 7 from 1. Regroup.

Step 3 Subtract the hundreds.

hundreds	tens	ones
5	11	11
6̸	2̸	1̸
− 4	7	5
1	4	6

Practice

Subtract.

1.

4	5	3
− 3	5	1

2.

6	1	2
− 1	5	9

Subtract.

3.

□	□	□
5	3	7
−	2	6

4.

□	□	□
6	3	4
− 2	7	8

5.

□	□	□
3	6	4
− 1	7	2

6.
```
  634
−  26
```

7.
```
  264
− 168
```

8.
```
  524
− 445
```

9.
```
  347
− 168
```

10. 344 people are at a game. 198 people leave at half time. How many people are still at the game?

_____ people

Test Practice

11. Which problem needs to be regrouped twice?

354 − 134 ◯

367 − 263 ◯

364 − 274 ◯

364 − 278 ◯

Touchdown!

Math at Home Have your child explain how regrouping tens and hundreds are the same.

See and Show

You can rewrite a problem to subtract.
Find 368 − 279.

Step 1 Rewrite.

Step 2 Subtract.

$$
\begin{array}{r}
2\;|15|18 \\
\cancel{3}\;\cancel{6}\;\cancel{8} \\
-\;2\;\;7\;\;9 \\
\hline
8\;\;9
\end{array}
$$

Line up the ones, tens, and hundreds.

Rewrite the problem. Subtract.

1. 336 − 272

2. 377 − 264

3. 633 − 265

4. 264 − 175

5. 845 − 378

6. 555 − 428

Talk Math How is rewriting three-digit subtraction different than when you rewrite two-digit subtraction?

On My Own

Rewrite the problem. Subtract.

7. 363 − 278

8. 285 − 185

9. 634 − 175

10. 375 − 142

11. 825 − 195

12. 647 − 373

13. 695 − 295

14. 853 − 259

15. 496 − 349

16. 495 − 267

17. 845 − 264

18. 764 − 375

Problem Solving

19. Lucas is reading a book that is 239 pages long. He has read 159 pages. How many pages does he have left to read?

_____ pages

20. Our class read 753 books over the summer. The girls read 394 books. How many books did the boys read?

I do a lot of summer reading!

_____ books

21. Our school library has 125 books about pets. I have read 96 of the books. How many books do I have left to read?

_____ books

 Explain how you rewrite three-digit subtraction problems to solve.

_____

_____

_____

Name _____

My Homework

Homework Helper **Need help?** connectED.mcgraw-hill.com

Find 356 − 298.

Step 1 Rewrite. Place the greater
number on top.

Step 2 Subtract. Regroup if necessary.

$$\begin{array}{c|c|c}
2 & 14 & 16 \\
\cancel{3} & \cancel{5} & \cancel{6} \\
- \quad 2 & 9 & 8 \\
\hline
& 5 & 8 \\
\end{array}$$

Practice

Rewrite the problem. Subtract.

1. 724 − 235

2. 616 − 337

3. 374 − 286

4. 875 − 596

5. 945 − 387

6. 435 − 294

Rewrite the problem. Subtract.

7. 162 − 89

8. 619 − 254

9. 195 − 99

10. 835 people are on a boat. 295 people get off of the boat. How many people are still on the boat?

Ahoy!

_____ people

Test Practice

11. Which problem shows how to rewrite the problem?

368 − 179 = _____

368	179	368	368
− 179	− 368	− 179	+ 179
200	411	189	547
◯	◯	◯	◯

 Math at Home Write a three-digit subtraction number sentence for your child. Have him or her rewrite the number sentence vertically and then subtract.

Problem Solving
STRATEGY: Write a Number Sentence

Fido always begs for treats. Watch
He has a bag of 255 treats.
By the end of the month, he had
eaten 82 treats. How many treats
are left for Fido to eat?

Tasty!

1 Understand <u>Underline</u> what you know.
Circle what you need to find.

2 Plan How will I solve the problem?

3 Solve Write a number sentence.

$$255 - 82 = 173$$ treats are left

4 Check Is my answer reasonable? Explain.

Practice the Strategy

385 students entered the science fair. 193 of the students were girls. How many students were boys?

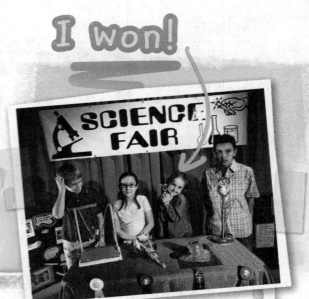

I won!

SCIENCE FAIR

1 Understand Underline what you know. Circle what you need to find.

2 Plan How will I solve the problem?

3 Solve I will...

_____ – _____ = _____

4 Check Is my answer reasonable? Explain.

Apply the Strategy

1. It took 227 days to build railroad tracks over a mountain. It took 132 days to build tracks over flat ground. How many more days did it take to build over the mountain?

_____ ◯ _____ ◯ _____ days

2. The Fuller family is driving 475 miles. They have already gone 218 miles. How many miles are left to go?

_____ ◯ _____ ◯ _____ miles

3. Flora's class is trying to collect 850 cans. She has turned in 370 cans. How many cans are needed to reach the class goal?

_____ ◯ _____ ◯ _____ cans

Review the Strategies

Choose a strategy
- Make a model.
- Guess, check, and revise.
- Write a number sentence.

4. Anna has been keeping track of the weather for 289 days. 196 of the days were sunny. How many days were not sunny?

_____ days

5. There are 836 flowers in the field. The children pick 398 of the flowers. How many flowers are left in the field?

_____ flowers

6. 423 fish are in the tank. 184 fish are removed from the tank. How many fish are in the tank now?

_____ fish

Name _____

My Homework

638 people started the race. 459 people finished the race. How many people did not finish the race?

1 Understand Underline what you know.
Circle what you need to find.

2 Plan How will I solve the problem?

3 Solve Write a number sentence.

638 – 459 = 179

people started   people finished   people did not finish

179 people did not finish the race.

4 Check Is my answer reasonable?

Underline what you know. Circle what you need to find.

1. A farmer picks 389 ears of corn.
   He sells 183 ears of corn. How many
   ears of corn does the farmer have left?

   _____ ◯ _____ ◯ _____ ears of corn

2. 135 people are at the park.
   46 people are playing baseball.
   How many people are not playing baseball?

   _____ ◯ _____ ◯ _____ people

3. There are 276 animals at the zoo.
   185 of the animals are male.
   How many animals are female?

   _____ ◯ _____ ◯ _____ animals

Test Practice

4. 152 tomatoes grew in my garden last year.
   This year only 98 tomatoes grew. How many
   more tomatoes grew last year?

252	250	54	52
◯	◯	◯	◯

Math at Home Have your child find 477 − 293 by writing a number sentence.

Name ....

Subtract Across Zeros

Explore and Explain

Yippee, I'm a mammal!

Me too!

_____ animals

 Teacher Directions: Model using base-ten blocks. We have learned about 200 different kinds of animals. 126 of those animals were mammals. How many of the animals were not mammals? Draw the blocks. Write the number.

See and Show

 Processes & Practices

You can subtract across zeros.

Find 400 − 234.

Step 1 Subtract the ones. You cannot subtract 4 ones from 0 ones. There are no tens to subtract from. Look at the hundreds. Regroup 1 hundred as 10 tens. Then regroup 1 ten as ten ones.

hundreds	tens	ones
□ 4̶ 2	□ 0̶ 3	10 0̶ 4
		6

Step 2 Subtract the tens. There are 9 tens left. Subtract 3 tens from 9 tens.

hundreds	tens	ones
□ 4̶ 2	9 0̶ 3	10 0̶ 4
	6	6

Step 3 Subtract the hundreds. There are 3 hundreds left. Subtract 2 hundreds from 3 hundreds.

400 − 234 = __166__

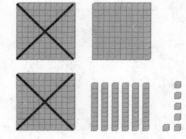

hundreds	tens	ones
3 4̶ 2	9 0̶ 3	10 0̶ 4
1	6	6

Talk Math How is subtracting from 400 different than subtracting from 435?

Name _____

On My Own

Use Work Mat 7 and base-ten blocks. Subtract.

1.

hundreds	tens	ones
□	□	□
8	0	0
− 5	3	2

2.

hundreds	tens	ones
□	□	□
7	0	0
− 6	1	4

3.
$$\begin{array}{r} 100 \\ -\ 76 \\ \hline \end{array}$$

4.
$$\begin{array}{r} 900 \\ -\ 287 \\ \hline \end{array}$$

5.
$$\begin{array}{r} 400 \\ -\ 167 \\ \hline \end{array}$$

6.
$$\begin{array}{r} 700 \\ -\ 444 \\ \hline \end{array}$$

7.
$$\begin{array}{r} 300 \\ -\ 16 \\ \hline \end{array}$$

8.
$$\begin{array}{r} 800 \\ -\ 477 \\ \hline \end{array}$$

9.
$$\begin{array}{r} 900 \\ -\ 876 \\ \hline \end{array}$$

10.
$$\begin{array}{r} 500 \\ -\ 54 \\ \hline \end{array}$$

11.
$$\begin{array}{r} 800 \\ -\ 691 \\ \hline \end{array}$$

12.
$$\begin{array}{r} 200 \\ -\ 75 \\ \hline \end{array}$$

13.
$$\begin{array}{r} 500 \\ -\ 321 \\ \hline \end{array}$$

14.
$$\begin{array}{r} 600 \\ -\ 312 \\ \hline \end{array}$$

Problem Solving

15. 400 bees are in the hive. 145 bees leave the hive. How many bees are still in the hive?

_____ bees

16. 300 girls took gymnastics last year. This year 193 girls take gymnastics. How many more girls took gymnastics last year than this year?

_____ girls

HOT Problem 500 people signed up for soccer. 123 people stopped playing. Then 154 more people stopped playing. How many people still play soccer? Explain.

_____

_____

_____

Name

My Homework

Homework Helper **Need help?** connectED.mcgraw-hill.com

You can subtract across zeros.

Find 600 − 336.

Step 1 Subtract the ones.
Regroup 1 hundred
into 10 tens. Then
regroup 1 ten
as 10 ones.

hundreds	tens	ones
5	9	10
6̸	0̸	0̸
− 3	3	6
2	6	4

Step 2 Subtract the tens.

Step 3 Subtract the hundreds.

Practice

Subtract.

1.
$$\begin{array}{r} 300 \\ -\ 251 \\ \hline \end{array}$$

2.
$$\begin{array}{r} 600 \\ -\ 139 \\ \hline \end{array}$$

3.
$$\begin{array}{r} 700 \\ -\ 386 \\ \hline \end{array}$$

4.
$$\begin{array}{r} 200 \\ -\ 126 \\ \hline \end{array}$$

5.
$$\begin{array}{r} 500 \\ -\ 385 \\ \hline \end{array}$$

6.
$$\begin{array}{r} 800 \\ -\ 272 \\ \hline \end{array}$$

Subtract.

7. 100 − 89

8. 600 − 564

9. 500 − 268

10. 900
− 432

11. 700
− 364

12. 200
− 147

13. 400 pineapples are planted in a row. 293 pineapples get picked. How many pineapples are still in the row?

_____ pineapples

I spiked my hair!

Test Practice

14. Find 600 − 289.

889   311   301   489

○   ○   ○   ○

Math at Home Have your child explain how to subtract 392 from 800.

Name _____

My Review

Vocabulary Check

place value   regroup   subtract

Write the correct word in the blank.

1. To identify the value of the digit 9 in the numbers 976, 93, and 9, use what you know about _____.

2. You can _____ to find the difference.

3. To take a number apart and write it a new way is to _____.

Concept Check

Subtract.

4. 600 − 400 = _____

5. 700 − 600 = _____

Hooray, we're free!

Subtract.

6.
$$
\begin{array}{r}
800 \\
- 400 \\
\hline
\end{array}
$$

7.
$$
\begin{array}{r}
900 \\
- 600 \\
\hline
\end{array}
$$

8.
$$
\begin{array}{r}
773 \\
- 100 \\
\hline
\end{array}
$$

9.
$$
\begin{array}{r}
261 \\
- 10 \\
\hline
\end{array}
$$

10.
$$
\begin{array}{r}
938 \\
- 329 \\
\hline
\end{array}
$$

11.
$$
\begin{array}{r}
885 \\
- 16 \\
\hline
\end{array}
$$

12.
$$
\begin{array}{r}
357 \\
- 189 \\
\hline
\end{array}
$$

13.
$$
\begin{array}{r}
987 \\
- 598 \\
\hline
\end{array}
$$

14.
$$
\begin{array}{r}
201 \\
- 124 \\
\hline
\end{array}
$$

Rewrite the problem. Subtract.

15. $385 - 166$

16. $247 - 189$

17. $925 - 638$

Subtract.

18.
$$
\begin{array}{r}
400 \\
- 254 \\
\hline
\end{array}
$$

19.
$$
\begin{array}{r}
700 \\
- 443 \\
\hline
\end{array}
$$

20.
$$
\begin{array}{r}
300 \\
- 165 \\
\hline
\end{array}
$$

Name _____

21. There are 620 buttons. 200 of them are square. The rest are round. How many buttons are round?

_____ buttons

22. Clara had 359 pennies. She lost some. Now she has 266 pennies. How many pennies did Clara lose?

_____ pennies

Test Practice

23. There are 334 cars in the parking lot on Sunday. On Monday, there are 182 cars. How many more cars were there on Sunday?

334	182	152	150
○	○	○	○

Reflect

Show the ways to subtract three-digit
numbers.

Subtract hundreds. 800 − 500 = _____	**Regroup tens to subtract.** 835 − 726
Regroup tens and hundreds to subtract. 935 − 397	**Rewrite to subtract.** 381 − 298

ESSENTIAL QUESTION

How can I subtract
three-digit numbers?

You are part
of this equation!

You + School = Cool

Glossary/Glosario

Vocab abc ← Go online for the eGlossary.

English	Spanish/Español
A.M. The hours from midnight until noon.	**a.m.** Las horas que van desde la medianoche hasta el mediodía.
add (addition) Join together sets to find the total or sum. The opposite of *subtract*. $$2 + 5 = 7$$	**sumar (adición)** Unir conjuntos para hallar el total o la suma. Lo opuesto de restar. $$2 + 5 = 7$$
addend Any numbers or quantities being added together. In $2 + 3 = 5$, 2 is an addend and 3 is an addend. $$2 + 3 = 5$$	**sumando** Numeros o cantidades que se suman. En $2 + 3 = 5$, 2 es un sumando y 3 es un sumando. $$2 + 3 = 5$$

after Follow in place or time.

5 6 7 8

6 is just *after* 5

después Que sigue en lugar o en tiempo.

5 6 7 8

6 está justo *después* del 5

analog clock A clock that has an hour hand and a minute hand.

minute hand — hour hand

reloj analógico Reloj que tiene una manecilla horaria y un minutero.

minutero — manecilla horaria

angle Two sides on a two-dimensional shape meet to form an angle.

ángulo Dos lados de una figura bidimensional se encuentran para formar un ángulo.

array Objects displayed in rows and columns.

row →

column

arreglo Objetos organizados en filas y columnas.

fila →

columna

Bb

bar graph A graph that uses bars to show data.

gráfica de barras Gráfica que usa barras para ilustrar datos.

before

5 **6** 7 8

6 is just *before* 7

antes

5 **6** 7 8

6 está justo *antes* del 7

between

47 48 **49** 50

49 is *between* 48 and 50

entre

47 48 **49** 50

49 está *entre* 48 y 50

Cc

cent

1¢   1 cent

centavo

1¢   1 centavo

Cc

cent sign (¢) The sign used to show cents.

1¢   5¢

signo de centavo (¢) El signo que se usa para mostrar centavos.

1¢   5¢

centimeter (cm) A metric unit for measuring length.

centímetro (cm) Unidad métrica para medir la longitud.

circle A closed, round two-dimensional shape.

círculo Bidimensional redonda y cerrada.

compare Look at objects, shapes, or numbers and see how they are alike or different.

comparar Observar objetos, formas o números para saber en qué se parecen y en qué se diferencian.

cone A three-dimensional shape that narrows to a point from a circular face.

cono Figura tridimensional que se estrecha hasta un punto desde una base circular.

count back On a number line, start at the greater number (5) and count back (3).

$$5 - 3 = 2$$

2 3 4 **5** 6

contar hacia atrás En una recta numérica, comienza en un número mayor (5) y cuenta (3) hacia atrás.

$$5 - 3 = 2$$

2 3 4 **5** 6

count on On a number line, start at the first addend (4) and count on (2).

$$4 + 2 = 6$$

3 4 5 6 7

seguir contando En una recta numérica, comienza en el primer sumando (4) y cuenta (2) hacia delante.

$$4 + 2 = 6$$

3 4 5 6 7

Cc

cube A three-dimensional shape with 6 square faces.

cubo Figura tridimensional con 6 caras cuadradas.

cylinder A three-dimensional shape that is shaped like a can.

cilindro Figura tridimensional que tiena la forma de una lata.

Dd

data Numbers or symbols, sometimes collected from a survey or experiment, that show information. *Data* is plural.

Name	Number of Pets
Mary	3
James	l
Alonzo	4

datos Números o símbolos que se recopilan mediante una encuesta o experimento para mostrar información.

Nombre	Número de mascotas
María	3
James	l
Alonzo	4

day I day = 24 hours
Examples: Sunday,
Monday, Tuesday,
Wednesday, Thursday,
Friday, Saturday

día I día = 24 horas
Ejemplos: domingo, lunes,
martes, miércoles, jueves,
viernes y sábado

difference The answer to a subtraction problem.

$$3 - 1 = 2$$
↑
The difference is 2.

diferencia Resultado de un problema de resta.

$$3 - 1 = 2$$
↑
La diferencia es 2.

digit A symbol used to write numbers. The ten digits are

0, 1, 2, 3, 4, 5, 6, 7, 8, 9.

dígito Símbolo que se utiliza para escribir números. Los diez dígitos son
0, 1, 2, 3, 4, 5, 6, 7, 8, 9.

digital clock A clock that uses only numbers to show time.

reloj digital Reloj que marca la hora solo con números.

Dd

---

dime dime = 10¢ or 10 cents

head　　tail

moneda de 10¢ moneda de diez centavos = 10¢ o 10 centavos

cara　　cruz

---

dollar one dollar = 100¢ or 100 cents. It can also be written as $1.00 or $1.

front

back

dólar un dólar = 100¢ o 100 centavos. También se puede escribir $1.00.

frente

revés

---

dollar sign ($) The sign used to show dollars.

one dollar = $1 or $1.00

signo de dólar ($) Símbolo que se usa para mostrar dólares.

un dólar = $1 o $1.00

---

doubles (and near doubles) Two addends that are the same number.

6 + 6 = 12 ← doubles

6 + 7 = 13 ← near doubles

dobles (y casi dobles) Dos sumandos que son el mismo número.

6 + 6 = 12 ← dobles

6 + 7 = 13 ← casi dobles

---

edge The line segment where two *faces* of a three-dimensional shape meet.

edge →

arista Segmento de recta donde se encuentran dos caras de una figura tridimensional.

arista →

equal groups Each group has the same number of objects.

There are four equal groups of counters.

grupos iguales Cada grupo tiene el mismo número de objetos.

Hay cuatro grupos iguales de fichas.

equal parts Each part is the same size.

This sandwich is cut into 2 equal parts.

partes iguales Cada parte es del mismo tamaño.

Este sándwich está cortado en 2 partes iguales.

Ee

equal to =

$$6 = 6$$

6 is *equal to* or the same as 6.

igual a =

$$6 = 6$$

6 es *igual* o lo mismo que 6.

---

estimate Find a number close to an exact amount.

47 + 22 rounds to 50 + 20.
The estimate is 70.

estimar Hallar un número cercano a la cantidad exacta.

47 + 22 se redondea a 50 + 20.
La estimación es 70.

---

even number Numbers that end with 0, 2, 4, 6, 8.

número par Los números que terminan en 0, 2, 4, 6, 8.

---

expanded form
The representation of a number as a sum that shows the value of each digit. Sometimes called *expanded notation*.

536 is written as 500 + 30 + 6.

forma desarrollada
La representación de un número como suma que muestra el valor de cada dígito. También se llama *notación desarrollada*.

536 se escribe como 500 + 30 + 6.

face The flat part of a three-dimensional shape.

A square is a face of a cube.

cara La parte plana de una figura tridimensional.

El cuadrado es la cara de un cubo.

fact family Addition and subtraction sentences that use the same numbers.

$$6 + 7 = 13 \qquad 13 - 7 = 6$$
$$7 + 6 = 13 \qquad 13 - 6 = 7$$

familia de operaciones Enunciados de suma y resta los cuales tienen los mismos números.

$$6 + 7 = 13 \qquad 13 - 7 = 6$$
$$7 + 6 = 13 \qquad 13 - 6 = 7$$

foot (ft) A customary unit for measuring length. Plural is feet.

1 foot = 12 inches

pie Unidad usual para medir longitud.

1 pie = 12 pulgadas

fourths Four equal parts of a whole. Each part is a fourth, or a quarter of the whole.

cuartos Cuatro partes iguales de un todo. Cada parte es un cuarto, o la cuarta parte del todo.

Gg

greater than >

7 > 2

7 is greater than 2.

mayor que >

7 > 2

7 es mayor que 2.

group A set of objects.

1 group of 4

grupo Conjunto o grupo de objetos.

1 grupo de 4

half hour (or half past)
A unit to measure time.
Sometimes called *half past*
or *half past the hour*.

a half hour = 30 minutes

media hora (o y media)
Unidad para medir tiempo.
A veces se dice *hora y
media*.

media hora = 30 minutos

halves   Two equal parts of
a whole. Each part is a half
of the whole.

mitades   Dos partes iguales
de un todo. Cada parte es
la mitad de un todo.

hexagon   A 2-dimensional
shape that has six sides.

hexágono   Figura
bidimensional que tiene seis
lados.

hour A unit to measure time.

1 hour = 60 minutes

hora Unidad para medir tiempo.

1 hora = 60 minutos

hour hand The hand on a clock that tells the hour. It is the shorter hand.

hour hand

manecilla horaria Manecilla del reloj que indica la hora. Es la más corta.

manecilla horaria

hundreds The numbers in the range of 100–999. It is the place value of a number.

365

3 is in the hundreds place.
6 is in the tens place.
5 is in the ones place.

centenas Los números en el rango de 100 a 999. Es el valor posicional de un número.

365

3 está en el lugar de las centenas.
6 está en el lugar de las decenas.
5 está en el lugar de las unidades.

inch (in) A customary unit for measuring length. The plural is *inches*.

12 inches = 1 foot

pulgada (pulg) Unidad usual para medir longitud.

12 pulgadas = 1 pie

inverse Operations that are opposite of each other.

Addition and subtraction are inverse or opposite operations.

operaciones inversas Operaciones que se oponen una a otra.

La suma y la resta son operaciones inversas u opuestas.

Kk

key Tells what (or how many) each symbol stands for.

Favorite Pet

Fish	☺	☺	☺	
Dog	☺			
Cat	☺	☺		

Key: ☺ = 2 votes

clave Indica qué o cuánto representa cada símbolo.

Animal doméstico favorito

Pez	☺	☺	☺	
Perro	☺			
Gato	☺	☺		

Clave: ☺ = 2 votos

length How long or how far away something is.

length

longitud El largo de algo o lo lejos que está.

longitud

less than <

4 < 7

4 is less than 7.

menor que <

4 < 7

4 es menor que 7.

line plot A graph that shows how often a certain number occurs in data.

diagrama lineal Una gráfica que muestra con qué frecuencia ocurre cierto número en los datos.

measure To find the length, height, weight, capacity, or temperature using standard or nonstandard units.

medir Hallar la longitud, altura, peso, capacidad o temperatura usando unidades estándares o no estándares.

meter (m) A metric unit for measuring length.

1 meter = 100 centimeters

metro (m) Unidad métrica para medir longitud.

1 metro = 100 centímetros

minute (min) A unit to measure time. Each tick mark is one minute.

1 minute = 60 seconds

minuto (min) Unidad para medir tiempo. Cada marca es un minuto.

1 minuto = 60 segundos

minute hand The longer hand on a clock that tells the minutes.

minute hand

minutero La manecilla más larga del reloj. Indica los minutos.

minutero

Mm

missing addend The missing number in a number sentence that makes the number sentence true.

$$9 + \boxed{} = 16$$

The missing addend is 7.

sumando desconocido El número desconocido en un enunciado numérico que hace que este sea verdadero.

$$9 + \boxed{} = 16$$

El sumando desconocido es 7.

month A unit of time.
12 months = 1 year

April						
Sunday	Monday	Tuesday	Wednesday	Thursday	Friday	Saturday
		1	2	3	4	5
6	7	8	9	10	11	12
13	14	15	16	17	18	19
20	21	22	23	24	25	26
27	28	29	30			

This is the month of April.

mes Unidad de tiempo.
12 mesas = 1 año

Abril						
domingo	lunes	martes	miércoles	jueves	viernes	sábado
		1	2	3	4	5
6	7	8	9	10	11	12
13	14	15	16	17	18	19
20	21	22	23	24	25	26
27	28	29	30			

Este es el mes de abril.

Nn

near doubles Addition facts in which one addend is exactly 1 more or 1 less than the other addend.

casi dobles Operaciones de suma en las cuales un sumando es exactamente 1 más o 1 menos que el otro sumando.

nickel nickel = 5¢ or 5 cents

head   tail

moneda de 5¢ moneda de cinco centavos = 5¢ o 5 centavos

cara   cruz

number line A line with number labels.

recta numérica Recta con marcas de números.

o'clock At the beginning of the hour.

It is 7 o'clock.

en punto El momento en que comienza cada hora.

Son las 7 en punto.

odd number Numbers that end with 1, 3, 5, 7, 9.

números impares Los números que terminan en 1, 3, 5, 7, 9.

Copyright © The McGraw-Hill Companies, Inc.   United States coin images from the United States Mint

Oo

ones The numbers in the range of 0-9. A place value of a number.

65

5 is in the ones place.

unidades Los números en el rango de 0 a 9. Valor posicional de un número.

65

El 5 está en el lugar de las unidades.

---

order

1, 3, 6, 8, 10

These numbers are in order from least to greatest.

orden

1, 3, 6, 8, 10

Estos números están en orden de menor a mayor.

Pp

P.M. The hours from noon until midnight.

p.m. Las horas que van desde el mediodía hasta la medianoche.

---

parallelogram A two-dimensional shape that has four sides. Each pair of opposite sides is equal and parallel.

paralelogramo Figura bidimensional que tiene cuatro lados. Cada par de lados opuestos son iguales y paralelos.

---

partition To divide or "break up."

separar Dividir o desunir.

pattern An order that a set of objects or numbers follows over and over.

pattern unit

patrón Orden que sigue continuamente un conjunto de objetos o números insertar punto.

unidad de patrón

penny penny = 1¢ or 1 cent

head   tail

moneda de 1¢ moneda de un centavo = 1¢ o 1 centavo

cara   escudo

pentagon A polygon with five sides.

pentágono Polígono de cinco lados.

picture graph A graph that has different pictures to show information collected.

gráfica con imágenes Gráfica que tiene diferentes imágenes para ilustrar la información recopilada.

Cómo voy a la escuela								
En Autobús								
En Bicicleta								
Caminando								

place value The value given to a *digit* by its place in a number.

1,365

1 is in the thousands place.
3 is in the hundreds place.
6 is in the tens place.
5 is in the ones place.

valor posicional El valor dado a un *dígito* según su posición en un número.

1,365

1 está en el lugar de los millares.
3 está en el lugar de las centenas.
6 está en el lugar de las decenas.
5 está en el lugar de las unidades.

pyramid A three-dimensional shape with a polygon as a base and other faces that are triangles.

pirámide Figura tridimensional con un polígono como base y otras caras que son triángulos.

Qq

quadrilateral A shape that has 4 sides and 4 angles.

cuadrilátero Figura con 4 lados y 4 ángulos.

quarter quarter = 25¢ or 25 cents

head   tail

moneda de 25¢ moneda de 25 centavos = 25¢ o 25 centavos

cara   cruz

quarter hour A quarter hour is 15 minutes. Sometimes called *quarter past* or *quarter til*.

cuarto de hora Un cuarto de hora es 15 minutos. A veces se dice hora y cuarto.

rectangle A plane shape with four sides and four corners.

rectángulo Figura plana con cuatro lados y cuatro esquinas.

Rr

rectangular prism A three-dimensional shape with 6 faces that are rectangles. 	**prisma rectangular** Figura tridimensional con 6 caras que son rectángulos.
regroup Take apart a number to write it in a new way. I ten + 2 ones becomes 12 ones.	**reagrupar** Separar un número para escribirlo de una nueva forma. I decena + 2 unidades se convierten en 12 unidades.
related fact(s) Basic facts using the same numbers. Sometimes called a *fact family*. 4 + 1 = 5 5 − 4 = 1 1 + 4 = 5 5 − 1 = 4	**operaciones relacionadas** Operaciones básicas en las que se usan los mismos números. También se llaman *familias de operaciones*. 4 + 1 = 5 5 − 4 = 1 1 + 4 = 5 5 − 1 = 4

repeated addition To use the same addend over and over.

suma repetida Usar el mismo sumando una y otra vez.

rhombus A shape with 4 sides of the same length.

rombo Paralelogramo con cuatro lados de la misma longitud.

round Change the *value* of a number to one that is easier to work with.

24 rounded to the nearest ten is 20.

redondear Cambiar el *valor* de un número a uno con el que es más fácil trabajar.

24 redondeado a la decena más cercana es 20.

side One of the line segments that make up a shape.

A pentagon has five sides.

lado Uno de los segmentos de recta que componen una figura.

El pentágono tiene cinco lados.

skip count Count objects in equal groups of two or more.

2, 4, 6, 8, 10

contar salteado Contar objetos en grupos iguales de dos o más.

2, 4, 6, 8, 10

slide To move a shape in any direction to another place.

slide

deslizar Trasladar una figura a una nueva posición.

deslizar

sphere A three-dimensional shape that has the shape of a round ball.

esfera Figura tridimensional que tiene la forma de una pelota redonda.

square A two-dimensional shape that has four equal sides. Also a rectangle.

cuadrado Figura bidimensional que tiene cuatro lados iguales. También es un rectángulo.

subtract (subtraction) Take away, take apart, separate, or find the difference between two sets. The opposite of *add*.

$$5 - 5 = 0$$

restar (sustracción) Eliminar, quitar, separar o hallar la diferencia entre dos conjuntos. Lo opuesto de *sumar*.

$$5 - 5 = 0$$

sum The answer to an addition problem.

$$2 + 4 = 6$$

suma Resultado de la operación de sumar.

$$2 + 4 = 6$$

survey Collect data by asking people the same question.

Favorite Animal	
Dog	⊬⊬⊬ I
Cat	⊬⊬⊬

This survey shows favorite animals.

encuesta Recopilar datos al hacer las mismas preguntas a un grupo de personas.

Animal favorito	
Perro	⊬⊬⊬ I
Gato	⊬⊬⊬

Esta encuesta muestra los animales favoritos.

symbol A letter or figure that stands for something.

This symbol means to add.

símbolo Letra o figura que representa algo.

Este símbolo significa sumar.

Tt

tally marks A mark used to record data collected in a survey.

⊬⊬⊬ II

tally marks

marca de conteo Marca que se utiliza para registrar los datos recopilados en una encuesta.

⊬⊬⊬ II

marcas de conteo

tens The numbers in the range of 10–99. It is the place value of a number.

65

6 is in the tens place.
5 is in the ones place.

decenas Los números en el rango de 10 a 99. Es el valor posicional de un número.

65

6 está en el lugar de las decenas.
5 está en el lugar de las unidades.

thirds Three equal parts.

tercios Tres partes iguales.

thousand(s) The numbers in the range of 1,000–9,999. It is the place value of a number.

1,365

1 is in the thousands place.
3 is in the hundreds place.
6 is in the tens place.
5 is in the ones place.

millar(es) Los números en el rango de 1,000 a 9,999. Es el valor posicional de un número.

1,365

1 está en el lugar de los millares.
3 está en el lugar de las centenas.
6 está en el lugar de las decenas.
5 está en el lugar de las unidades.

three-dimensional shape A shape that has length, width, and height.

figura tridimensional Que tiene tres dimensiones: largo, ancho y alto.

trapezoid A two-dimensional shape with four sides and only two opposite sides that are parallel.

trapecio Figura bidimensional de cuatro lados con solo dos lados opuestos que son paralelos.

triangle A three-dimensional shape with three sides and three angles.

triángulo Figura tridimensional con tres lados y tres ángulos.

two-dimensional shape The outline of a shape - such as a triangle, square, or rectangle - that has only *length* and *width*.

figura bidimensional Contorno de una figura, como un triángulo, un cuadrado o un rectángulo, que solo tiene *largo* y *ancho*.

vertex

vértice

week A part of a calendar.
I week = 7 days

semana Parte de un calendario
una semana = 7 días

whole The entire amount or object.

el todo La cantidad total o el objeto completo.

yard (yd) A customary unit for measuring length.

I yard = 3 feet or 36 inches

yarda Unidad usual para medir la longitud.

I yarda = 3 pies o 36 pulgadas

year A way to count how much time has passed or will pass. I year = 12 months

año Un período insertar punto, un año = 12 meses

January

S	M	T	W	T	F	S
						1
2	3	4	5	6	7	8
9	10	11	12	13	14	15
16	17	18	19	20	21	22
23	24	25	26	27	28	29
30	31					

February

S	M	T	W	T	F	S
		1	2	3	4	5
6	7	8	9	10	11	12
13	14	15	16	17	18	19
20	21	22	23	24	25	26
27	28					

March

S	M	T	W	T	F	S
		1	2	3	4	5
6	7	8	9	10	11	12
13	14	15	16	17	18	19
20	21	22	23	24	25	26
27	28	29	30	31		

April

S	M	T	W	T	F	S
					1	2
3	4	5	6	7	8	9
10	11	12	13	14	15	16
17	18	19	20	21	22	23
24	25	26	27	28	29	30

May

S	M	T	W	T	F	S
1	2	3	4	5	6	7
8	9	10	11	12	13	14
15	16	17	18	19	20	21
22	23	24	25	26	27	28
29	30	31				

June

S	M	T	W	T	F	S
			1	2	3	4
5	6	7	8	9	10	11
12	13	14	15	16	17	18
19	20	21	22	23	24	25
26	27	28	29	30		

July

S	M	T	W	T	F	S
					1	2
3	4	5	6	7	8	9
10	11	12	13	14	15	16
17	18	19	20	21	22	23
24	25	26	27	28	29	30
31						

August

S	M	T	W	T	F	S
	1	2	3	4	5	6
7	8	9	10	11	12	13
14	15	16	17	18	19	20
21	22	23	24	25	26	27
28	29	30	31			

September

S	M	T	W	T	F	S
				1	2	3
4	5	6	7	8	9	10
11	12	13	14	15	16	17
18	19	20	21	22	23	24
25	26	27	28	29	30	

October

S	M	T	W	T	F	S
						1
2	3	4	5	6	7	8
9	10	11	12	13	14	15
16	17	18	19	20	21	22
23	24	25	26	27	28	29
30	31					

November

S	M	T	W	T	F	S
		1	2	3	4	5
6	7	8	9	10	11	12
13	14	15	16	17	18	19
20	21	22	23	24	25	26
27	28	29	30			

December

S	M	T	W	T	F	S
				1	2	3
4	5	6	7	8	9	10
11	12	13	14	15	16	17
18	19	20	21	22	23	24
25	26	27	28	29	30	31

enero

D	L	M	M	J	V	S
						1
2	3	4	5	6	7	8
9	10	11	12	13	14	15
16	17	18	19	20	21	22
23	24	25	26	27	28	29
30	31					

febrero

D	L	M	M	J	V	S
		1	2	3	4	5
6	7	8	9	10	11	12
13	14	15	16	17	18	19
20	21	22	23	24	25	26
27	28					

marzo

D	L	M	M	J	V	S
		1	2	3	4	5
6	7	8	9	10	11	12
13	14	15	16	17	18	19
20	21	22	23	24	25	26
27	28	29	30	31		

abril

D	L	M	M	J	V	S
					1	2
3	4	5	6	7	8	9
10	11	12	13	14	15	16
17	18	19	20	21	22	23
24	25	26	27	28	29	30

mayo

D	L	M	M	J	V	S
1	2	3	4	5	6	7
8	9	10	11	12	13	14
15	16	17	18	19	20	21
22	23	24	25	26	27	28
29	30	31				

junio

D	L	M	M	J	V	S
			1	2	3	4
5	6	7	8	9	10	11
12	13	14	15	16	17	18
19	20	21	22	23	24	25
26	27	28	29	30		

julio

D	L	M	M	J	V	S
					1	2
3	4	5	6	7	8	9
10	11	12	13	14	15	16
17	18	19	20	21	22	23
24	25	26	27	28	29	30
31						

agosto

D	L	M	M	J	V	S
	1	2	3	4	5	6
7	8	9	10	11	12	13
14	15	16	17	18	19	20
21	22	23	24	25	26	27
28	29	30	31			

septiembre

D	L	M	M	J	V	S
				1	2	3
4	5	6	7	8	9	10
11	12	13	14	15	16	17
18	19	20	21	22	23	24
25	26	27	28	29	30	

octubre

D	L	M	M	J	V	S
						1
2	3	4	5	6	7	8
9	10	11	12	13	14	15
16	17	18	19	20	21	22
23	24	25	26	27	28	29
30	31					

noviembre

D	L	M	M	J	V	S
		1	2	3	4	5
6	7	8	9	10	11	12
13	14	15	16	17	18	19
20	21	22	23	24	25	26
27	28	29	30			

diciembre

D	L	M	M	J	V	S
				1	2	3
4	5	6	7	8	9	10
11	12	13	14	15	16	17
18	19	20	21	22	23	24
25	26	27	28	29	30	31

Name _____

Work Mat 1: Ten-Frame

Work Mat 2: Ten-Frames

Work Mat 2: Ten-Frames

Name

Work Mat 3: Number Lines

Work Mat 4: Number Lines

Work Mat 4: Number Lines

Name

Work Mat 5: Part-Part-Whole

Part	Part

Whole

Tens	Ones

Copyright © The McGraw-Hill Companies, Inc.

Work Mat 6: Tens and Ones Chart

Work Mat 6: Tens and Ones Chart

Name _____

Work Mat 7: Hundreds, Tens, and Ones Chart

Hundreds	Tens	Ones

Work Mat 8: Thousands, Hundreds, Tens, and Ones Chart

Thousands	Hundreds	Tens	Ones

Work Mat 8: Thousands, Hundreds, Tens, and Ones Chart